BERRIES

A HARROWSMITH GARDENER'S GUIDE

BERRIES

EDITED BY JENNIFER BENNETT
ILLUSTRATIONS BY MARTA SCYTHES

CAMDEN HOUSE

CAMDEN
•HOUSE•
PUBLISHING

Canadian Cataloguing in Publication Data

Main entry under title:

Berries: a Harrowsmith gardener's guide
ISBN 0-921820-19-4

1. Berries. I. Bennett, Jennifer.

SB381.B47 1991   634'.7   C91-093133-X

Trade distribution by
Firefly Books
250 Sparks Avenue
Willowdale, Ontario
Canada M2H 2S4

Printed and bound in Canada by
D.W. Friesen & Sons Ltd.
Altona, Manitoba, for
Camden House Publishing
(a division of Telemedia Publishing Inc.)
7 Queen Victoria Road
Camden East, Ontario
K0K 1J0

Design by
Linda J. Menyes

Cover by
Ron Broda

Colour separations by
Hadwen Graphics
Ottawa, Ontario

Printed on acid-free paper

Acknowledgements

Berries: A Harrowsmith Gardener's Guide represents, in its finished state, the commitment and cooperation of many individuals. They include art director Linda Menyes; artist Marta Scythes, who prepared the illustrations; editor Tracy C. Read; assistant editor Mary Patton; photo researcher Jane Good; publishing coordinator Mirielle Keeling; typesetter Patricia Denard-Hinch; production manager Susan Dickinson; copy editor Catherine DeLury; and associates Ellen Brooks Mortfield, Jane Crawford, Charlotte Du-Chene, Laura Elston, Christine Kulyk and Christina Tracy.

Contents

Introduction

In her 1881 *Studies of Plant Life in Canada*, Catharine Parr Traill described more than a dozen types of berries that grew within a few miles of her family's small farm in what were then the wilds of Upper Canada. She expressed a preference for several species of blueberry: "These pretty shrubs, loaded with their luscious berries, may be found on all dry, open places. The poor Indian squaw fills her bark baskets with the fruit and brings them to the villages to trade for flour, tea and calico, while social parties of the settlers used to go forth annually to gather the fruit for preserving or for the pleasure of spending a long summer's day among the romantic hills and valleys, roaming in unrestrained freedom among the wildflowers that are scattered in rich profusion over those open tracts of land where these useful berries grow. These rural parties would sometimes muster to the extent of 50 or even 100 individuals, furnished with provisions and all the appliances for an extended picnic." ∞ Whether harvested for necessity or enjoyment, wild berries have always been available to us. One of the authors of this gardener's guide, botanist Nancy J. Turner, is an expert on wild berries of North America and is particularly interested in their use by indigenous people. Bunchberries, for instance, have been used extensively by the natives of the West Coast, though few more recent settlers consider the plants anything other than pretty little wild things. Catharine Parr Traill, too, was a keen observer of the people who were intimately acquainted with the local wild fruits of the area long before Europeans arrived. The native people taught her how to recognize the plants and how to use them, although her palate did not always agree with theirs. Of partridgeberry, she wrote: "The berries are mealy and insipid but are eaten by the Indian women and children as a dainty." Traill picked elderberries, gooseberries, currants, Juneberries, huckleberries, blueberries, cranberries, raspberries, blackberries, sheepberries, partridgeberries and the fruit of hawthorns and wild roses. That legacy of enjoying nature's bounty is carried on by Jo Ann Gardner, who writes about the collection and preparation of berries, wild and cultivated, in Chapter Five.

Cultivated berries are far newer fruits, the results of human selection and hybridization. Some, such as strawberries, have been grown in gardens for centuries. The first popular gardening book published in the English language, in 1577, recorded that "certain skilful men, by a diligence and care, procure the berries to alter from the proper red colour to a fair white, delectable to the eye." Since then, skilful plant lovers have produced cultivated sorts of other wild berries, and they have created new hybrids, too, such as tayberries, boysenberries, loganberries, day-neutral strawberries and many more, all described on the following pages.

Domestic berries are developed from wild plants but usually have larger fruit that may also be sweeter, more prolific and easier to pick. The price for larger fruit that is just outside the back door instead of a half-hour's hike away is, of course, cultivation. Fruit that you grow yourself requires care, though no more care than a perennial flowerbed. A decided bonus is that, like flowerbeds, berry bushes can be decorative in the home landscape, where they were habitually grown before this century.

In the 1911 book *Landscape Gardening as Applied to Home Decoration*, the author, Samuel T. Maynard, includes berries and fruit trees along with the usual lawns, larches and pathways. Including useful plants among the decorative ones was a carry-over from pioneer days, when time was too limited for much of it to be spent coddling inessentials.

Maynard listed blackberries, raspberries, currants, gooseberries and strawberries as small fruits that the homeowner would simply not be without. Catharine Parr Traill added to that list the highbush cranberry – "As a garden shrub, this viburnum is considered very ornamental, from its abundance of flowers and beautiful fruit" – and wondered why more was not made of hawthorns: "There appears to have been little attempt made to cultivate our hawthorns as hedge plants." She would be satisfied to know that some plant nurseries now sell both hawthorns and highbush cranberries.

It should be mentioned at this point that not all of these fruits are true berries in the botanical sense. Nancy J. Turner, author of Chapter Four, writes, "Technically, these fruits are classed in different categories. Highbush cranberries and bunchberries, having a single seed or stone surrounded by fleshy tissue, are known as drupes, like cherries, peaches and plums. Saskatoon berries are classified as pomes, like apples and pears. Raspberries, blackberries and strawberries are known as aggregate fruits. Finally, there are the true berries, including blueberries, currants and wild grapes, which have many small seeds scattered within soft, fleshy tissue and are usually globular in shape." In this book, the word "berry" is used to refer to any small fleshy fruits. However, wild cherries and plums (*Prunus* spp) and their relatives have been excluded because their domesticated counterparts are not generally considered berries.

Wild berries are a delectable part of human history. Bunchberries (Cornus canadensis) *were extensively eaten, fresh or cooked, by the indigenous people of North America.*

The once common type of landscaping that involves useful plants is now rare enough to have its own name – edible landscaping, or permaculture – which is practised not so much because homeowners have limited time but because they appreciate the higher quality of homegrown, freshly picked fruit. A backyard supply not only means fresh fruit in season but can also mean produce grown without chemicals that may be harmful. In this book, all the authors have concentrated on organic techniques that are healthy for plants, soil and people.

In the landscape, these small fruits can be grown in a patch on their own – the best method for raspberries and other brambles or cane fruits, described in Chapter Three – or they can be kept within the vegetable garden or even the flowerbed, in line with the principles of edible landscaping. The berries described in Chapter One are normally grown in vegetable gardens, either because they are annual types that allow the ground to be turned after harvest or because, like strawberries, they need ongoing maintenance. Bush fruits, however, are particularly capable of assuming the role of ornamental shrubs beside the house, edging the driveway or in flower borders, where they may become the tallest design elements. There must be easy access for picking and pruning, however. They are described in Chapter Two.

Before any planting takes place, the gardener must be familiar with the plants: their needs, habits and potential height and breadth. Catharine Parr Traill writes, "Every plant, flower and tree has a simple history of its own, not without its interest if we would read it aright. It forms a page in the great volume of nature which lies open before us – in nature, there is no space left unoccupied."

–Jennifer Bennett 11

Chapter One:
Three for the Row

By Jennifer Bennett

F rom June's first strawberries through July's blueberries and August's gooseberries until the ripening of the last of the fall-bearing raspberries, berries of various types mark the passage of brief but bright northern summers in a most delicious way. For gardeners, berries are by and large an easy lot – not very demanding of space or time, not as likely to be troubled by diseases as fruit trees – yet they are generous producers of predictable crops. For cooks, fresh garden or wild berries are treasures. Many types cannot be bought in the supermarket for any price. Others deteriorate so fast that only the luckiest shoppers find high-quality fresh fruit. For the landscape architect, berries offer ornamental plants suited to almost any spot that is not entirely in shade. And for those of us who remember secret childhood forays into forbidden adult gardens, berries evoke a simple yet powerful sense of nostalgia. ∽ Strawberries, ground cherries and garden huckleberries are fruits for containers, baskets, vegetable gardens and even flowerbeds. Because all of them are herbaceous plants that die back to the ground in winter, they are versatile berries that can be moved from place to place from year to year or skipped entirely should the gardener feel a pressing need for an extra row of peas or onions. I have, at one time or another, grown all of these movable berries in my vegetable garden and 13 found all of them easy and productive, although not all of them are necessarily worth repeating.

Garden Huckleberry

The garden huckleberry, or black nightshade (*Solanum nigrum*), is the only one I would not grow again. It is a peculiar fruit that bears no relationship – botanical or gustatory – to the delicious huckleberry of the West Coast (*Vaccinium parvifolium*). I had such fond memories of those "real" huckleberries that, in my first garden in eastern Ontario, I turned to the garden huckleberry as a possible substitute.

I started seeds indoors in six-packs on March 26 and set plants outdoors on May 28, after the last spring frost. Small white flowers appeared about a month later, and by midsummer, the black berries, larger than blueberries, were ready to eat. Unfortunately, they tasted mealy and unappealing. The garden huckleberry is related to the potato, but it more closely resembles another relative, deadly nightshade (*Atropa belladonna*), with its 2- or 3-foot self-supporting bushes yielding clusters of dusky berries that turn almost black when fully ripe. There are those who claim that, bolstered with enough sugar, garden huckleberries make a fine pie – the plant has some dedicated fans – but almost any fruit will make a palatable dessert when enough sugar is added to it. In any case, the plant is so productive, barring a season-long drought, that many a garden dabbler might want to give it a try. Each plant produces about a quart of fruit – enough for one sugary pie. Experimenters, however, should taste the fruit only when it is fully ripe and cooked, as the immature raw fruit contains a poisonous alkaloid.

The famous plant breeder Luther Burbank reported that he had created a hybrid of two *Solanum* species, which he called sunberry and someone else dubbed wonderberry. Almost from the first, how-

Garden huckleberries should not be eaten until they are fully ripe and cooked.

ever, there were doubters who claimed that the wonderberry was simply black nightshade given a new name. More likely, Burbank's fruit was another species of edible nightshade, but as his notes are confusing, the truth may never be known.

Ground Cherry

Another member of the nightshade family – one I found much tastier – is the ground cherry (*Physalis peruviana*), whose little yellow berries grow inside a husk in the manner of its relatives, the Chinese lantern (*P. alkekengi*) and the tomatillo (*P. ixocarpa*). The genus name *Physalis* comes from the Greek word for bladder on account of the characteristic husk, formed from the flower's calyx, which has enlarged to cover the berry. The tomatillo, a favourite Mexican vegetable since prehistoric times, produces savoury yellowish green berries an inch or more across.

Started indoors in the manner of tomatoes, they can be grown quite easily in any northern garden and used in authentic Mexican dishes. I made a simple sauce from my own tomatillo crop and froze it to use in the winter.

The ground cherry, also called cape gooseberry, husk tomato and strawberry tomato, is much sweeter. There are several species (see Chapter Four), but the usual one sold by garden-seed companies is another South American native. I started seeds indoors April 22 and set them 2 feet apart in a sunny place in good soil on May 26. Sprawling plants grew a couple of feet tall, producing small yellow flowers followed by round fruits that began to ripen to yellow-orange by mid-August while the husks turned brown. When the half-inch fruits are ripe, they fall to the ground. Harvested and left in the husks in a cool place indoors, the berries will keep for several weeks. Remove the husks before eating ground cherries fresh or cooked in jams, pies and other desserts. Ground cherries often self-sow, provided some fruits have been left on the ground.

Strawberries

If the first two movable berries are little known, strawberries are North America's most popular fruit. The two fruits described above are annuals, growing from seed to fruit in a single growing season, but strawberries are perennials. A gardener who obtains a few healthy plants and sets them in good garden soil in a sunny place need never buy strawberry plants again – or the fruit either, for that matter: 25 standard strawberry plants, well cared for, will yield 15 to 30 quarts of berries a year. It is not enough, however, to buy a few plants, stick them in the ground and expect fruit forever.

Strawberry plants must be treated well at the outset and every year thereafter. Twice, I have planted strawberries in places where the roots of perennial weeds

The yellow fruits of the ground cherry form inside a distinctive papery husk.

still lurked underground, and within a couple of months, the plants could scarcely be seen through a tangle of twitch grass and horsetails. They reluctantly yielded just a few berries that were plucked off by marauding groundhogs. Choose your best garden soil for strawberries. One gardener I know laid down photodegradable plastic mulch on tilled garden rows and then inserted the plants through holes in the plastic. The system worked beautifully. Late the same summer, the plastic was entirely hidden under lush strawberry foliage, and enough daughter plants were growing beyond the edges of the plastic that the procedure could be repeated in an adjoining row the next year.

There are three types of standard runner-bearing strawberries: June-bearing, everbearing and day-neutral. (Runners are the stems that grow out along the ground, producing new plants at their tips.) The first and most common is the 15

Strawberries must be planted on weed-free soil, where they can renew themselves in-definitely by putting forth runners, stems that grow along the ground.

June-bearing strawberry. In these plants, the long, warm days of late spring and early summer trigger the growth of runners, while the cooler temperatures and shorter days of late summer and early fall urge flower buds to form. All berries on these plants develop from buds formed the previous year. The plants produce a single big crop of berries in June – or later, farther north – and are labelled early, midseason or late, depending upon when the crop ripens during the month.

June Bearers

June bearers, the standard berries for the north, are generous, hardy, dependable and easy to propagate. Some of the most cold-hardy cultivars are:

'Blomidon,' from Nova Scotia, is a late-flowering favourite in northern Canada. The fruit is large but difficult to hull. Plants are susceptible to mildew.

'Bounty,' from Nova Scotia, produces heavy crops of delicious, late-season, medium red berries.

'Glooscap,' from Nova Scotia, is extremely cold-tolerant, so it has done well in the Prairie Provinces, New Brunswick and Minnesota. Plants, yields and midseason fruits are all large.

'Guardian,' from Maryland, produces midseason berries that are slightly larger than those of 'Redcoat.' Plants are vigorous and resistant to red stele and verticillium wilt.

'Honeoye,' from New York, yields large, bright red fruit that tends to darken when overripe. The plants produce runners freely and are very high-yielding, but they are susceptible to red stele and verticillium wilt.

'Kent,' from Nova Scotia, is an extremely high-yielding, midseason berry with vigorous plants that produce abundant runners. In trials in Illinois, Indiana,

Iowa, Michigan, Minnesota, Missouri, Ohio and Wisconsin between 1980 and 1988, 'Kent' averaged the highest yield of all 15 strawberries tested. (Second was 'Honeoye.') The foliage is slightly susceptible to mildew. The large, glossy fruit is red throughout but rather difficult to hull.

'Protem,' from Alberta, is very hardy, the best choice for places too cold for any of the others. The fruit has good flavour.

'Redchief,' a traditional midseason berry from Maryland, is resistant to red stele, verticillium wilt, leaf scorch and mildew. The berries are medium-sized, tasty and deep red. Later berries have a slight neck at the top.

'Redcoat,' an Ontario release and one of the most popular commercial berries in the province, is a very heavy early midseason producer of berries whose flavour is considered only fair. Strong, winter-hardy plants produce runners freely.

'Sparkle,' also called 'Paymaster,' is a 1942 release from New Jersey. The crimson berries are soft but delicious. Vigorous plants produce runners excessively, and while they are resistant to powdery mildew, they are susceptible to verticillium wilt.

'Veestar,' from Ontario, is an early berry that equals 'Redcoat' in productivity and has better flavour. Plants are vigorous and produce runners well, although they are susceptible to leaf scorch and mildew.

Everbearers

The everbearing, or fall-bearing, strawberry is newer. It yields the usual spring berries, then blooms again in summer to produce a fall crop from about August until October or the first hard frost. A few everbearing strawberries are available from northern catalogues:

'Fort Laramie' bears large, bright red fruit that maintains its size well in cool, moist areas. It will tolerate winter temperatures of minus 30 degrees F.

'Ogallala,' from Wyoming, is a very hardy producer of tender berries that become slightly bitter in hot weather.

'Ozark Beauty,' also known as 'Cross's Red' and 'Autumn Beauty,' is an Arkansas release that produces large, wedge-shaped berries with red flesh and prominent yellow seeds. Plants produce runners moderately well.

Day-Neutrals

Newer still and most exciting are the day-neutral strawberries, which produce flower buds all season, regardless of day length. A large June crop is followed by a short rest and then fruit until frost. Because the day-neutrals generally have better flavour and fruit quality than the everbearers, they have taken the limelight as alternatives to June bearers. Day-neutrals also greatly outyield June bearers in their first year – as much as 10 times the number of berries in one Saskatchewan trial – and even in their second year, the per-acre weight of day-neutral crops was three times that of June-bearing cultivars in Ontario trials. Furthermore, where late spring frosts are common, their summer blooming increases the likelihood of at least some crop.

On the negative side, day-neutral berries are often a bit smaller and the plants are less hardy than the hardiest June bearers. A crop spread over a longer season can be a shortcoming if you want berries in quantity to freeze or make jam out of when few other garden crops are demanding attention. Day-neutrals are more susceptible to pest damage, do best with watering and fertilizing all summer – fish emulsion or a topdressing of compost or well-rotted manure every two weeks – and produce runners less prolifically, so it is not easy to rejuvenate or increase a patch. Andrew Jamieson, a small-fruit breeder for the Agriculture Canada Research Station in Kentville, Nova Scotia, suggests renewing a day-neutral plot with new plants every two years. 17

'Hecker' is a California release that does well on the prairies, where it bears early enough to produce a full crop before autumn frosts.

'Tribute,' from Maryland, is resistant to red stele and verticillium wilt. The fruit is firm and generally larger than that of 'Tristar.' The prefix "tri" in both names indicates the ability to set fruit in spring, summer and fall. 'Tribute' bears its three crops somewhat later than 'Tristar.'

'Tristar,' a sister of 'Tribute,' produces smallish plants that produce runners fairly well, especially if early blossoms are removed. It yields a heavy spring crop of delicious, deep red, medium-sized fruit that is better tasting than that of 'Tribute.' Plants are resistant to red stele, verticillium wilt, leaf scorch and leaf blight.

Whether you choose June-bearing, everbearing or day-neutral strawberries, the planting process will be much the same. Choose a sunny place sheltered from strong winds, and avoid low-lying areas, where frost damage is most likely to occur. Mix into the soil a couple of inches of compost or well-rotted manure to provide the sort of rich, loose, well-drained bed strawberries like best. A green-manure crop planted and tilled under the year before will also increase soil fertility and help control weeds.

Alpines

While most strawberries are purchased as plants and, indeed, will not come true from seed, some selections of wild strawberries can be grown from seed. These produce small, delicious berries that may crop, as mine do, both spring and fall. Although I never pick much fruit from my 2-by-4-foot bed of seed-grown alpine strawberries, there are often a few berries ready for picking as I walk past. 'Sweetheart' is a relatively large-fruited selection but is still smaller than a standard berry. My plants do not produce runners but will self-sow if fruit falls on the ground.

Wild strawberries planted about 6 inches apart create a ground cover.

To grow alpine strawberries, start the seeds indoors in January or February, covering them lightly. Seedlings should appear in three or four weeks if temperatures are between 55 and 75 degrees F, after which the pots must be moved to a sunny windowsill, into a cold frame or under plant lights. When daytime temperatures reach the 60s, plant the strawberries outdoors about 6 inches apart.

Plant runner strawberries soon after the ground can be worked in spring. When purchasing plants, look for healthy ones with lots of pale roots; discard those with black roots. Do not let the roots either dry out or become waterlogged; dip them into a pail of cool water before planting. Set the plants into the soil at the same depth at which they grew before, with roots entirely buried and the midpoint of the crown at ground level. Space June-bearing plants 1½ to 2 feet apart down the row, and allow runners to fill the empty spaces between the plants. Everbearers and day-neutrals produce runners less prolifically, so they should be planted about 12 and 6 inches apart, respectively. They can be staggered in double or triple rows, with paths between the rows. In subsequent years, remove all older plants directly after fruiting, leaving only young plants spaced about 6 inches apart.

After planting, remove all the blossoms from June-bearing plants to encourage the growth of strong roots and plenty of buds for the next year's crop. On everbearers and day-neutrals, pinch off the flowers just until the end of June so that there will be a first crop in late summer. In following years, blossoms are not normally pruned.

Water immediately after planting, as the fine roots can dry out quickly. At least half of a strawberry's roots grow in the top 6 inches of soil, so if plants do not receive about an inch of water a week between blossoming and harvest, small, dry berries will result. University of Missouri researchers found that watering plants between midsummer and early autumn was also essential for large berries, because that is when the cell size of the fruit bud is determined. Properly timed watering, they discovered, could increase the next year's yields by as much as 5,000 quarts per acre. Drip irrigation – perforated tubes laid on the ground and connected to an outdoor tap – is an effective way to use relatively little water but is expensive and time-consuming to install. In a small planting, most gardeners prefer to hand-water. If sprinkling, aim the water at the soil rather than the foliage, and water on a still evening for greatest efficiency.

Mulching in fall, which helps protect roots and crowns from frost damage, is a good idea everywhere except in the mildest areas of the West Coast and the south. After a few light frosts but before the end of November, cover the plants with evergreen branches or weed-free straw 3 or 4 inches deep. One bale will cover about 100 square feet. Tradition has it that the

mulch should be removed in spring as soon as signs of new growth appear and the strawberry foliage turns light yellow, usually around late April. In experiments at Cornell University, however, earlier mulch removal consistently enhanced yields and accelerated flowering. Remove the mulch in late winter to allow the plants to finish developing their flower buds in the sunlight. The mulch can be tucked between plants or spread in the paths. Be prepared to re-cover the plants with straw, newspaper or blankets if frost threatens after blossoming has begun; freezing will lower the harvest. Spunbonded plastic mulches such as Reemay have been used successfully to extend the

The small, sweet fruits of some wild strawberries can be grown from seed.

fall harvest of day-neutrals and as winter mulches in areas where temperatures do not drop below 20 degrees F. These translucent covers will also speed the warming process in spring. Leave them in place until flowering begins.

As with other crops, the diseases that harm strawberries cause more problems for commercial growers than for home gardeners, in part because the latter are more forgiving of small crops and imperfect fruit. Nevertheless, buying plants from a reputable nursery and purchasing 19

Although there are everbearing and day-neutral strawberries, the most common types are June bearers such as the large-fruited, high-yielding 'Honeoye.'

stock that has inbred resistance to diseases troublesome in your area are good practices. Another is to move the strawberry bed every two or three years, avoiding places planted within the last three years with tomatoes, eggplants, peppers, potatoes or raspberries, all of which can carry the fungal disease verticillium wilt. Symptoms of this fatal disease usually appear first during late summer of the planting year: outer leaves wilt and turn reddish with upturned margins, and stems develop black areas.

Additional fungal diseases include leaf scorch, which can defoliate entire plantings, and black root rot, which is aggravated by extremely cold winters; mulching will help prevent it. Grey mould is self-descriptive. It is the most serious disease of day-neutrals, because berries are present on the plant continuously throughout the growing season. Remove mouldy berries every day, and dispose of them away from the garden. Powdery mildew worsens in hot, dry weather, overwinters in infected plant debris and causes leaves to curl and die. Infected plants can be treated with a wettable sulphur powder applied after harvest and in spring when growth starts. Red stele, another fungal disease, causes wilting and plant death. It is identifiable by a reddish core, or stele, in the roots and is aggravated by cool, wet soil. All fungal diseases are lessened if plants are adequately spaced and beds are kept clean of weeds and old and dead plants. The growth of misshapen berries (known as catfacing) is not caused by disease but by incomplete pollination, which can result from plant damage by pests, from temperatures that are too high or too low or from a scarcity of honeybees at blossoming time.

Harvesttime is about three to five weeks after blossoming, depending mostly upon weather and cultivar. Pick every day or two, pinching off the stems with the strawberries; removal of the hulls when picking greatly accelerates spoilage. Refrigerate the berries as soon as possible, and use them within a few days. To freeze strawberries, place them whole or sliced on cookie sheets, freeze them until firm, then pack them into airtight containers, or carefully mix half a cup of sugar into every cup of sliced or whole berries, then freeze. Berries frozen with sugar retain more flavour, firmness and colour than those frozen without.

Renovate the strawberry bed as soon as the harvest is finished. Remove the foliage with shears or a lawn mower set at its highest setting, being careful not to damage the crowns. Dig out any weeds, and thin the plants, removing all the older ones. Mix an inch or two of compost or well-rotted manure and a dusting of bone meal into the soil between plants.

Cooked, raw or frozen, strawberries offer not only gustatory benefits but medicinal ones as well. Their therapeutic qualities, celebrated for centuries, can be attributed in part to a very high ascorbic-acid content and in part to the fruit's undisputed value as a purgative. Carl Lin-

A new crop of blossoms is evidence of a bed renovated after last year's harvest.

naeus, who developed the modern system of plant taxonomy, swore by strawberries after they cured him of a severe attack of gout: "The disease returned the following year and also the year after but in an ever milder form, and the wild strawberries always cured me. Unfortunately, I cannot grow strawberries in winter, and I have wholly failed to preserve them." In sympathy with his plight, the queen of Sweden ordered that the fruit be grown for Linnaeus in the royal greenhouses.

Today, fresh strawberries in spring, when no other garden fruit is available, are just as valuable, their worth measured in the taste of sun, health and generosity.

Jennifer Bennett, who gardens in Ontario, is a senior contributing editor of *Harrowsmith* magazine, co-author of *The Harrowsmith Annual Garden* and author of *The Harrowsmith Northern Gardener* and *Lilies of the Hearth*.

Chapter Two:
In the Bush Garden

By Lewis Hill

Years ago, my family decided to raise as many fruits and berries as the Vermont climate, with its short, cool summers and frigid winters, would allow. Although we have had to forgo grapes and many of the common fruit trees, we can always count on harvesting generous crops of delicious blueberries, currants, elderberries, gooseberries, rose hips and saskatoons without scavenging in the wild for fruit during our busy summer. Currant jelly transforms ordinary toast into a winter-morning feast; tasty, vitamin-rich elderberry juice helps keep us healthy until spring; and all year, gooseberry pie rivals blueberry as my favourite food. ✑ Cold hardiness, then, is not the only reason we cultivate bush fruits. Most of all, we grow berry bushes because their fruit is delicious. By growing our own, we are sure of food that is free from harmful chemicals. Also, the plants, which are long-lived and attractive enough to be used in landscaping, are among the few things we can grow here that increase in value each year, while demanding only a small investment in time and money. They are troubled by few pests, and most of them produce fruit over a long season. ✑ All of this was well known to the pioneers a century ago, when bush fruits proved their reliability and hardiness to people who had little time to fuss over reluctant bearers. In our changing society, the cultivation of bush fruits has become 23 less popular as horticultural priorities have shifted. Now, most gardeners grow vegetables and flowers, but their blueberries come from commercial plots, their rose hips come in herbal tea bags, and their elderberries, currants and gooseberries are available only as pricey preserves in gourmet shops.

Starting Out

The beginner will be pleased to learn, however, that bush fruits are readily available and easy to plant. Many mail-order nurseries have a wide variety of bare-rooted plants for sale each spring. Local nurseries and garden centres often sell potted or balled plants. These are more expensive than the bare-rooted versions, but they can be planted anytime, will begin to grow faster in the garden and will probably bear heavily a year sooner. If healthy, however, either type of plant will be successful if given proper care. Most bush fruits grow well almost anywhere and in almost any type of soil. If a crop of lush grass grows well on your land, it is likely to be ideal for bush fruits too, although elderberries prefer soil that is a bit moist all year and blueberries need an especially acidic soil. Blueberries, saskatoons and roses thrive in a fairly light soil and full sun, but currants, elderberries and gooseberries can grow and produce well in somewhat heavier soil and with only half a day of sunlight.

A garden cliché worth repeating is that it is important to have your soil in good condition before you plant. A bit of preparation will not only start the plants off well but will spare you additional work and perhaps disappointment later. If you are doing a large planting, prepare the soil over the entire area as thoroughly as you would for a vegetable garden. For just a few bushes, however, it is easier and just as effective to plant each one as you would a fruit tree.

First, dig a hole at least twice the size

Bush fruits such as rose hips are easy to plant, and they grow well in almost any soil.

of the rootball. Divide the soil you removed from the hole in half, and into one half, mix an equal amount of compost and peat moss or well-aged manure. Put enough of the mixture into the hole to allow you to set the plant at the same depth that it grew originally. Fill the hole with water, then position the plant. Fill in the hole around the roots carefully with the remaining mix, tamping it lightly so that no air pockets will be left. Leave a slight depression in the soil around the base of the plant to catch the rain and future waterings. Now, to a pail of water add a small amount of liquid fertilizer such as manure tea, fish emulsion or liquid seaweed, and soak the soil around the plant thoroughly. Put the leftover soil on the compost heap, or scatter it between the plants on top of the ground.

A common mistake is to crowd fruit bushes too close together. A newly set infant plant looks pretty lonely 5 or 6 feet from its neighbour, but that is the minimum spacing for red and white currants and gooseberries. Black currants, elderberries, blueberries, saskatoons and rugosa roses spread even more and need at least 6 feet. Space rows of currants and gooseberries at least 7 feet apart and those of blueberries, saskatoons, elderberries, black currants and roses even farther, to leave room to walk between the rows when the plants are full grown. If you are trying to create a tight hedgerow of any of these bushes, you may choose to set the plants slightly closer together.

Most planting directions recommend two different varieties of each type of bush fruit for pollination. Currants and gooseberries, however, are self-fruitful, so will bear well even if only one kind is planted. These are good berries to choose, then, if you have only a small amount of space – in a city backyard garden, for example. Gardeners have found that more fruit is produced if they plant at least two varieties of elderberries or saskatoons; two varieties of blueberries are definitely needed.

We apply a layer of organic mulch around our new plants, extending it out about 2 feet from the stems. Mulch not only suppresses competing weeds and grasses but also helps prevent the moisture in the soil from evaporating, attracts earthworms and keeps the soil from becoming too warm in summer and from freezing hard in winter. Another bonus is that mulch adds fertilizer and humus to the soil as it rots. Last but not least, it protects the fruit on the lower branches from mud spatters in rainstorms. We use a 3-inch-deep layer of wood chips and bark peelings from a local mill, but many other materials make good mulches, including hay, grass clippings, shavings and leaves. Although we sometimes use newspapers or magazines, covering them with hay or chips for aesthetic reasons and to keep the papers from blowing away in the wind, we do not like to use sawdust – it packs too hard and locks up nitrogen where it touches the soil. Peat moss is not our favourite mulch either, because it tends to blow away when dry and it sheds water, so light rain and waterings do not soak into the soil.

After the plants are established and begin to grow in their new site, about the only care they need is an annual renewal of the mulch as it decomposes, additional fertilizer each spring and occasional light pruning. A few cups of dried manure or a layer of farm manure under each application of new mulch will furnish the necessary fertilizer. In late winter, we use a pair of hand pruners to cut out any broken or injured branches and all wood three or more years old.

Although bush fruits do not need to be picked every day in the harvest season as do raspberries and strawberries, they should not be left on the bush too long after they ripen. Some years, we were careless, waited too long, and the berries fell to the ground or were lost to the birds.

Blueberries

The cultivated blueberry is a comparative newcomer to the family of garden berries, although it has quickly become the most popular bush fruit for home gardens. We love to go out early in the morning to pick a few berries for our breakfast cereal.

The domesticated highbush blueberry, about 8 feet tall, was developed from the wild species *Vaccinium corymbosum*, which the early settlers found growing along the East Coast. The native people dried the fruit to store it for winter and showed the colonists how to do the same. Not until this century, however, did horticulturists make selections from the wild plants and crossbreed them. As a result of their research, more than 50 different large-fruited cultivars are now being grown.

Climate and soil affect blueberries more than any other bush fruits. Although hardy cultivars are being developed, most 25 of the early ones do not thrive in areas where winter temperatures dip much below minus 30 degrees F. Even in slightly warmer places, the plants are sensitive to the north wind and need to be grown in a protected spot. Hybrids of highbush and lowbush blueberries (*Vaccinium angustifolium*), however, extend the growing range to the prairies and north of the Great Lakes. These hybrids are not only hardier than the highbush types, but their dwarf size (2 or 3 feet) helps them survive because they are usually buried in snow for most of the winter. Some people prefer their slightly wild flavour too. Another highbush, the rabbit-eye (*V. ashei*), listed in some catalogues, is strictly for warmer, more southerly parts of the United States.

In the wild, highbush blueberries grow best in full sun on fairly light, very acidic peat soils, conditions the home gardener should try to imitate. Unlike those of other plants, blueberry roots do not produce the acid that enables them to absorb the nitrogen and iron necessary for growth, so they are unable to grow well unless the soil is extremely acidic, ranging in pH from 4.5 to 5. Any soil can be doctored enough to grow blueberries, but it may not always be worth the effort. One woman in our area sadly calculated that each berry on her bush had cost her a dollar. If your soil is naturally rich in lime, raising blueberries will require a lot of work, since the lime in the subsoil will percolate upward with each rain. A gardener with a soil that is only slightly alkaline can, however, make the soil acidic enough by adding 4 to 6 inches of sphagnum peat moss or pine needles to the top 6 to 8 inches of soil. Sulphur will also increase soil acidity. The Ontario Ministry of Agriculture and Food recommends adding to sandy loam soils 1½ to 2 pounds of sulphur per 100 square feet for each full point the soil registers above pH 4.5. For example, if the pH is 6.5, add 3 to 4 pounds per 100 square feet. (The soil's pH can be determined by a soil test.) For sandy soils, use about half that amount. Do not adjust the soil again for at least a year. After the bushes have been planted and watered, spread a 3-to-6-inch mulch of an acidic material such as pine needles, oak leaves, partially rotted wood chips or shavings of oak, pine or hemlock in a circle 2 feet wide around the plants. Mulching is preferable to cultivation because of the plant's shallow roots.

The roots of blueberries are very shallow compared with those of most other plants of the same size, and they have no root hairs to help them absorb water and fertilizer, so locate the plants away from shade trees, hedges or shrubs with aggressive roots that might rob the berries of water and nutrients. And because their shallow roots cannot reach deep into the earth for moisture, give the plants extra water during the growing and fruiting season—whenever nature does not provide. Feed them regularly by scattering nutrients on the soil surface. Blueberries need nitrogen more than phosphorus and potash, so include blood meal, fish emulsion, manure tea, soybean meal or cottonseed meal in the fertilizing ritual. Fertilize early in spring so that the plants will make their fastest growth in summer but slow down before the first killing frost. Northern growers should be careful not to encourage late growth.

Blueberry bushes grow more slowly than other small fruits, and they may require 10 years to reach full production, even though they may begin to bear soon after they are planted. In spite of the long wait, they are easy to care for in the right location and are not usually bothered by as many diseases as strawberries, raspberries and fruit trees. Start with at least two different varieties, three if possible, since cross-pollination is necessary for the plant to produce fruit and there is always a chance of losing one plant. A neighbour's bush or any wild blueberries can serve as pollinators if they bloom at the same time and grow within a few hundred feet of yours. Three or four bushes should pro-

Blueberry bushes grow more slowly than other small fruits and may take 10 years to reach full production, but they are easy to grow if they have sun, water and acidic soil.

duce enough fruit for a small family unless you intend to do a lot of preserving.

If you live in a cold-winter area, check with other local gardeners to see which varieties grow well. If at all possible, buy plants that were actually grown, not just sold, in a climate as severe as your own. They will already be acclimatized to short growing seasons and cold winters.

Of the highbush cultivars, the hardiest are 'Meader,' 'Northblue,' 'Northland' and 'Northsky.' Fairly hardy cultivars include 'Berkeley,' 'Bluecrop,' 'Blueray,' 'Jersey,' 'Patriot,' 'Rubel' and 'Weymouth.' The least hardy are 'Concord,' 'Coville,' 'Herbert' and 'Stanley.' Hardy lowbush cultivars include 'Augusta,' 'Brunswick' and 'Chignecto.'

As soon as the bushes begin to grow well, prune them every year or two in late winter, removing all branches that are injured, diseased, old or unproductive. Where winters are harsh, it is best not to prune blueberries for the first five or six years other than to remove winter damage. Train the plants to grow upright, and 27 thin out the twiggy ends of branches where the fruit was produced the previous season. Even after the plants begin to bear well, prune minimally so that you do not encourage rapid growth. Excessive new growth usually does not harden properly during the short growing season and thus is winter-killed.

Few insects and diseases bother the plants. The blueberry maggot and fruit fly are the most common insect pests, but they are unlikely to find a small backyard planting. If they do, rotenone will control them. Birds are the most common pests; the only protection is to cover the plants with netting well before the berries begin to ripen.

Blueberries are not easy to start from cuttings, and even well-established bushes do not sucker readily, which is why the plants are more expensive than other small fruits. The lower branches can be layered, but roots develop very slowly. Commercial growers now propagate most blueberries by tissue culture. If you want to expand your planting quickly, it is best to buy plants.

Currants and Gooseberries

Currants and gooseberries are relatively little known in North America, despite their popularity with immigrants from England, who discover that the large-fruited gooseberries they loved at home almost invariably develop mildew when grown here. Fortunately, the native species and their hybrids, though smaller-fruited, are very productive, more resistant to mildew and just as flavourful.

One of the reasons these plants of the genus *Ribes* have not been more widely grown in home gardens is their reputation for being the Typhoid Marys of the plant world. The plants are alternate hosts for white pine blister rust, a disease fatal to five-needled pines, including white pine and Swiss stone pine. The disease does not affect the *Ribes* themselves or

The European black currant is an alternate host of a disease fatal to some pines.

pines of other species such as red and Scotch pine.

Recent research has shown that blister rust is spread almost entirely by the native wild gooseberry, the wild currant and the European black currant (*Ribes nigrum*), which was brought to North America decades ago and has naturalized widely. Most horticulturists now believe that neither cultivated gooseberries nor cultivated currants are the villains and point out that plants have grown in the shade of white pines for many years without problems. Equally good news is that many of the new black currant cultivars are resistant to blister rust, and many have been planted in recent years.

But the argument is still far from settled. Canada has no laws prohibiting planting of cultivated *Ribes*, but many states require that currant and gooseberry bushes be planted a safe distance from white pine trees, usually 900 to 1,000 feet. Although the prohibition is rarely enforced, if you live in one of the states with such regulations, do think twice before planting *Ribes* near your neighbour's valuable stand of white pines.

Insects and diseases rarely affect the *Ribes* genus, but there are several pests that can strike when conditions are right.

Powdery mildew is one of the most common diseases to affect currants, especially black, and can be difficult to control where summers are warm and humid, although a sulphur spray may help. The best ways to prevent mildew are to plant currants in a sunny place with good air circulation and to keep the plants uncrowded. Anthracnose is a less common disease that shows up as leaf discolorations. *Ribes* may be attacked by leaf-eating currant or gooseberry worms in early summer, but these are easily checked with rotenone or a spraying of *Bacillus thuringiensis* (Bt).

A well-drained soil is best for black currants, but other currants and gooseberries prefer soil that is slightly heavy and loamy. All do best with a thick mulch over soil that is well supplied with organic matter and with a small amount of additional fertilizer each spring. The only pruning necessary on most *Ribes* is the removal of damaged branches and any wood more than three years old.

It is not difficult to start new *Ribes* plants from your own patch or a neighbour's in spring or fall. Simply dig up and transplant offshoots of the parent plant, or bend a branch to the ground and cover the middle portion with an inch of soil in early spring. Roots should form during the summer, and the following spring, cut the layered branch from the mother plant and transplant it. Currants can also be propagated from 6-inch hardwood cuttings taken in late winter. Gooseberries start better from softwood cuttings of the same length taken in early summer.

Currants

An attractive, 4-foot-tall currant bush drooping with red, white or black fruits is a savoury sight to any fruit lover. When the berries begin to ripen in midsummer, watch them carefully so that you will not miss picking them at their prime, or hungry birds will beat you to them. However, currants are at their best not when fresh

Red currants are best not when fresh but when cooked in desserts and preserves.

but after they have been cooked and the juice extracted for jelly, wine or syrup or when they have been made into conserves, pies, tarts, fools or numerous other desserts and condiments. Black currants are also tasty when dried, but the red and white kinds are too seedy. (The dried currants we buy in stores are small grapes.)

Ripe currants are not easy to pick, because they are small, soft and juicy. Although one can strip them from their stems on the bush, they will bruise less if the clusters are picked intact. Pick them from the stems when you are ready to process them. We usually harvest our entire crop of red and white currants at one picking but gather the black currants over a period of time, since they have a different flavour at each stage. Jo Ann Gardner, a well-known Nova Scotia fruit gardener and writer (and the author of Chapter Five), recommends picking the berries when they are slightly underripe 29

Gooseberries are so delicious that the author hides a few bushes from visitors.

resistant and rich in vitamin C. European black currants being tested in Canada and the United States include 'Ben Lomond,' 'Ben Nevis,' 'Brodtorp,' 'Ojebyn,' 'Rotkoop' and 'Titania.' Some are reputed to be superior to the Canadian types in quality, productivity and berry size.

'Jostaberry' is one of several gooseberry-black currant hybrids recently developed in Europe. It grows quickly, is said to be resistant to most of the *Ribes* pests, including blister rust, and produces fruits resembling gooseberries that are flavourful and rich in vitamin C. Another nice quality is the plant's lack of thorns. It is, however, somewhat more prone to winterkill than a currant. Where it is hardy, it is a heavy producer, so it should be pruned more heavily than other currants and gooseberries to prevent a decrease in the size of the berries.

Gooseberries

My wife and I pick gooseberries before they are completely ripe for tart pies, but we eat fully ripe ones straight off the bush. They are so delicious, in fact, that we learned to plant a few bushes out of sight of visitors so that we would have some berries left for cooking and winter storage. Gooseberries come in a great variety of sizes, and some plants are so thorny that we wear a heavy glove on one hand to hold the bush for easier picking.

Since the blister rust panic subsided, nurseries have begun to offer a better assortment of cultivars. Most grow 3 to 5 feet tall and, like currants, ripen their fruits in midsummer. Among the old types are the yellow-green, very thorny 'Downing,' the wine-red 'Poorman' and the green 'Oregon Champion,' which produces excellent fruit but is particularly good as a prickly hedge against wild garden invaders. 'Pixwell' – so named because its thorns are shorter than those of other gooseberries, which makes the fruit easier to pick – is the cultivar most often

for jelly and a bit riper for jam, because they are high in pectin at these stages. Ripe or overripe fruits can be used for juice and wine.

Pollination is not a problem for either currants or gooseberries, and a single plant will produce well by itself. For a small family, however, you will need two to four plants of each type of fruit if you plan to do any preserving.

Of the red currants, 'Cherry Red,' 'Red Lake' and 'Wilder' are most frequently offered. All are good, and there is very little difference between them. 'Cascade' has a sprawling type of growth.

The similar 'White Grape' and 'White Imperial' are the most popular white currants. The flavour of white currants is milder than that of the reds, and many people like to eat them fresh off the bush.

The most popular Canadian black currant cultivars are 'Consort,' 'Coronet,' 'Crusader' and 'Willoughby.' All are blight-

listed in catalogues. It has medium-sized green berries that turn pale pink when fully ripe. 'Fredonia' has large red berries on short, spiny plants. One of our favourites is 'Welcome,' not only because it has few thorns, but because it produces sweet, deep red fruits. Newer cultivars include the purple-fruited 'Captivator' and 'Leepared,' a Finnish cultivar that produces tasty, medium-sized, translucent red berries.

Vintage cookbooks advise readers to "top and tail" gooseberries when preparing them for jams, pies and other delights. We always remove the tough, pointed "tail" with a deft fingernail or a sharp knife before putting the berries in plastic bags to freeze for winter, but the process takes a lot of time, and some cooks swear that neither stem nor tail need be removed for most purposes.

Elderberries

When I first met the wild edible elder, I was not too impressed with it, because I had to pick a tremendous number of tiny berries to make even a small pie. When the hybrid elderberry cultivars began to appear in fruit catalogues, however, I decided to try them because they were described as cold-hardy and I hoped that we might add yet another berry to our collection of currants, gooseberries and bramble fruits.

We ordered four: 'Adams,' 'Nova,' 'New York 21' and 'York.' By chance, we planted them in an area that stays moist all summer. They grew rapidly, but the following year, we were surprised to find that our "tame" elderberries did not bloom until midsummer, later than their wild cousins. Consequently, they ripened after the blackberries. The birds found them the moment they ripened, of course, so we did not have enough to process that year, but we were encouraged. The following year, the clump grew about 8 feet tall, and we harvested a bushel of berries. Because of unusual weather that year, some of our

Like its cultivated cousin, the wild blue elderberry has a medicinal reputation.

songbirds had left the week before the fruit ripened, so we did not have to share.

'Adams' turned out to be the earliest, most vigorous and heaviest producer. Its red-black berries were larger than wild elderberries but still small. The other three produced much bigger berries, about the size of wild chokecherries. 'Nova,' a Canadian introduction, had delicious berries that ripened soon after 'Adams.' 'York' produced the largest fruit of the Canadian cultivars and bore heavily. 'New York 21,' the latest, never did ripen fully in our climate, so eventually, we dug them up. They are excellent fruits for regions with longer summers, however.

It did not take us long to discover that elderberries can be used for all the things that grapes are good for except eating fresh off the bush. We made them into juice, jelly and pies and experimented with a batch of wine. The elderberry flavour is rich and distinctive, but it is diffi- 31 cult to describe – perhaps a combination of grape, raspberry and blackberry.

In Europe, the elderberry has a fascinating history of use in medicine. Early botanists hailed the plant as the medicine of the common people, and the 17th-century Dutch physician and botanist Hermann Boerhaave was said to have tipped his hat in admiration every time he passed an elder tree. I, too, am a strong believer in the elderberry's health-giving qualities. Our family has had fewer colds and less flu each winter since we have been drinking the juice regularly. An elderly friend reports that her arthritis is much improved after daily drinks of an elderberry-orange juice mixture.

The fruit of Rosa rugosa *and other roses makes tasty teas, jellies and syrups.*

One cannot easily buy such an elixir over the counter, but elderberries are so easy to grow that there is no excuse for not having a patch. No special directions are necessary as long as they are planted in the right place at the outset. A friend put several new plants on a dry, sandy hillside, where they did not grow at all, but then moved them to a moist spot, where they grew profusely. Elderberries even do well on land that is too wet for blackberries. If you have sandy soil, apply plenty of compost, peat moss or well-rotted manure.

Mulch the bushes heavily, and water them often in dry weather.

In our harsh climate, the hybrids sometimes suffer winter damage, leaving the bushes unsightly until we prune off the dead limbs in late spring. The loss affects the crop very little, however, because the plants grow back vigorously and the fruit is produced on new growth.

We set elderberry plants about 5 feet apart and learned the hard way that rows should be at least 8 feet apart, because the plants spread quickly. Feed and mulch the young plants to start them off well, but after they mature, fertilize them only if they are not producing a good amount of new growth each year. The only pruning necessary is the removal of deadwood. We also find it useful to mow between the rows about once a week to keep sucker plants from spreading.

The only pests that bother our elderberry crop are birds. Unless we pick the berries the day before they are truly ripe, we lose most of them. Word of our cache apparently spreads quickly, causing every berry-eating bird in the township to stay north an extra week to stuff itself on the vitamin-rich Hill berries before it sets off on a healthy flight south. The birds not only strip off all the ripe berries but, even worse, break off entire clusters of unripe fruit by perching on them.

It is next to impossible to place bird netting over such tall bushes. We have tried hanging shiny aluminum plates or foil to no avail. We cope by picking the berries when they are still slightly green and letting them finish ripening in a fairly warm room. Within a day or two, they turn a deep purple-black and are ready to be stripped from their stems and made into pies or jelly or stewed into juice that we strain and freeze for winter. Last fall, we froze some juice in ice-cube trays and stored the cubes in plastic bags, ready to pop into a glass of fruit juice or cider.

The attractive elderberry plant, with its lush, tropical-looking foliage, is covered with huge umbels of tiny, fragrant, creamy white flowers in midsummer. Like our European ancestors, we eat the flower clusters dipped in batter and fried – a crisp, unusual hors d'oeuvre or dessert. Sometimes we place the freshly picked blossoms in a gallon glass jar, fill the jar with water to which a little sugar and lemon juice have been added and let the jar rest in the hot sun. In a few hours, we have a tasty, healthful sun tea.

By transplanting suckers from our half-dozen original plants, we now have rows of elderberries that extend more than 150 feet. One can also start new plants from cuttings or seeds, which sprout and grow to bearing size in two or three years, although the fruit from seedling plants varies considerably in quality.

Rose Hips

Even if it did not produce an abundance of vitamin C, a large *Rosa rugosa* bush that grows on the bank beside our greenhouse would be cherished. Not only are the large pink roses beautiful, but their sweet, spicy fragrance wafts throughout the yard. What follows – the attractive, orange-red fruit – is simply an additional feature. Rose fruits, called hips, are vitamin-packed bundles used to make tasty teas, jellies and syrups. In England and other northern countries during World War II, people who could not obtain their normal supply of citrus fruits took advantage of the high ascorbic acid content of rose hips by gathering tons to process and preserve for winter.

Many roses produce hips, but certain species and cultivars have particularly heavy yields. Most hips are marble-sized and difficult to pick, but one exceptionally fine producer is *Rosa rugosa*, an Asian species that bears huge fruits the size of crab apples. In some rural areas, this rose has naturalized, and pickers can find all the hips they want along roadsides and on abandoned farms. Native roses useful for

their hips are described in Chapter Four.

Both species and hybrid rugosas are available from most northern nurseries. The species has large, fragrant, pink or red flowers. Gardeners who prefer showy double flowers can choose from numerous hybrids, including 'Hansa,' 'Belle Poitevine,' 'Magnifica' and 'Rubra.' Although these varieties produce large hips, not all cultivars do. If you want hips, check the fine print of the catalogue or ask the nursery. Some of the new Canadian hardy roses are very good hip producers.

There are ways to obtain roses other than buying them. If you can find rugosas growing wild or suckers growing from a neighbour's bush, dig and transplant them when they are dormant in early spring or late fall. Rugosas also grow easily from seed, which can be planted either in flats or in the ground outdoors and transplanted when they are a foot high or taller. They grow well in almost any soil that is not too dry or too wet, but they do best if the land is at least moderately fertile. They need full sun to produce the best blooms and hips. Because rugosas are nearly wild, they will fend for themselves quite well and need little care. The plants usually flourish without additional fertilizer. Common rose pests such as Japanese beetles and rose chafers, which may bother them occasionally, can be controlled with rotenone.

The mature plants are 4 to 6 feet tall and, because they sucker profusely, must be restricted by frequent mowing unless you want a forest of rose bushes. (This trait, however, makes them useful for an impenetrable hedge, if one is wanted.) An annual cutting back when the plants are dormant will keep them from becoming too tall and floppy.

The hips are ripe when they turn a rich scarlet in autumn, but some people insist that the flavour is better after the first frost. Although a few aficionados eat hips directly off the bush, the texture of the fresh fruit is not appealing to us – lots of 33 seeds surrounded by pulp. We process them by stewing them into teas or jelly or making them into a syrup combined with honey or another sweetener. In Sweden, cold rose-hip soups are popular.

After picking, keep the fruits cool to help preserve the vitamins. Then, after washing, remove the blossoms and stem ends with scissors and cook the hips right away. Use enamel, stainless-steel or glass dishes, if possible, to avoid the vitamin loss that takes place in aluminum. One easy way to make syrup is to pour just enough boiling water over the hips to barely cover them and simmer for 15 to 20 minutes. Let the mixture cool and stand for a day before straining. Use the syrup in soups, dressings or toppings, or freeze it for later use. The hips can also be opened, scraped out and dried or frozen in bags until needed.

Saskatoons

The saskatoon is one of those wonderful marvels of nature that has adapted perfectly to a certain climate and fulfilled the needs of both the people and the wildlife living there. It is remarkable that such a plant not only can survive in the harsh expanses of the Northwest Territories, Prairie Provinces and northern states but also can produce tremendous crops.

Saskatoon fruits somewhat resemble blueberries in both appearance and flavour, although they are actually small pomes, like apples or pears. Most wild bushes produce fruits about a quarter of an inch in diameter or a little larger. The bush varies from a small shrub to 16 feet or taller, depending upon variety and growing conditions, and may produce 5 or 6 quarts of fruit. Commercial plantings have yielded as much as 6 tons per acre.

For saskatoons, choose a sunny spot with slightly acidic, well-drained soil with plenty of organic matter; saskatoons do not do well in heavy clay soil. The plants are very hardy, but their flowers bloom so early that they may be damaged by late spring frosts, so it is important to plant them where the air circulation is good and late spring frosts are not common. We have found that a thick mulch can hold winter's frost on the roots a bit longer, delaying blooming by a few days. Set the plants about 6 feet apart for a hedgerow, farther apart if you want space between the mature plants. If you plant more than one row, position the rows at least 10 feet apart. Keep the bushes weed-free by mulching, cultivating or mowing. If you cultivate, avoid deep tilling, which could damage the tender fibrous roots. Saskatoons usually begin to bear within three or four years.

Although wild saskatoons produce well with no pruning, both wild and cultivated plants do better if the old wood is removed from time to time. Avoid pruning when the wood is frozen. Prune after the coldest part of the winter is over but before the buds begin to swell. The fruit is produced on wood that grew the previous season, so severe cutback is not recommended.

New saskatoons can be started from seed, layering, offshoots or cuttings. The first three methods are easiest, since, as a rule, the plants start rather reluctantly from cuttings, either hardwood or softwood. Suckers or offshoots that grow from the roots can be carefully severed from the mother plant with a spade; if you are growing named plants grafted on wild roots, note that you will be propagating the wild species, not the named cultivar. To reproduce plants by layering, bend a lower branch to the ground and cover the middle portion with an inch of soil. After a year, the buried portion will have rooted and can be severed from the plant. Saskatoon seeds germinate readily, and a large percentage will produce plants that behave much like their parent, so for best results, collect seeds from plants that consistently produce good-sized crops of large berries with good flavour.

Birds love ripening saskatoons. Some

A cluster of ripening saskatoons promises some of the best berries that can be grown by prairie gardeners, whose winters are too harsh for more tender species.

commercial growers plant a hedgerow of wild saskatoons nearby, hoping the birds will have their fill of them and leave the later-ripening cultivars alone, but in our experience, this subterfuge does not help. Among insect pests, the most common are fruit maggots, which create wormy berries just as apple maggots produce wormy apples. Spraying soon after the fruit forms but before it is well developed helps control them. Use an environmentally benign spray that will wash off easily, such as insecticidal soap or rotenone.

The fruits, which grow in clusters like blueberries, are used much like blueberries and are delicious fresh or cooked in pies, jams and jellies. Pick the berries soon after they ripen: overripe fruit has less vitamin C and is not as good for preserving.

New saskatoon cultivars appear frequently. Most are selections made from seedling plants, but a few have resulted from careful hybridizing. 'Forestburg,' 'Pembina' and 'Smoky' are grown commercially in the Prairie Provinces. Of these, 'Forestburg' has the largest fruit, but its quality is not quite as good as that of 'Pembina' and 'Smoky.' 'Altaglow' produces a sweet berry on a bush that is also a fine ornamental. 'Honeywood' yields very large crops, and 'Moonlake' and 'Thiessen' have especially large fruits. 'Trim' and 'Success' are two recently developed cultivars.

Although saskatoons are of great importance to people who live where few fruit trees thrive and highbush blueberries are merely a dream, most people elsewhere either know little about them or find the wild berries too seedy to enjoy. This is partly because the large-fruiting named cultivars have not always done well beyond the prairies, although the University of Vermont maintains a successful planting of several kinds at their experimental station in Burlington. For the most part, saskatoons are popular fruits only where the better-known highbush blueberries are not hardy.

Lewis Hill, author of *Cold Climate Gardening* and *Fruits and Berries for the Home Garden*, operates a nursery in Greensboro, Vermont, with his wife Nancy.

Chapter Three:
Raising Canes

By Dorothy Hinshaw Patent

Bramble fruits became special to me when, at the age of 12, I was assigned the important but pleasant task of picking fat, juicy boysenberries from well-trained vines so that my mother could make what I thought were the best pies in the world. Until I moved to Montana and discovered wild huckleberries, no pie I tasted equalled the tangy-sweet pastries my mother had so lovingly and expertly produced. ✑ It was in Montana that, in my 30s, I first had my own garden. Boysenberries were at the top of my list of plants to grow. Unfortunately, I was naïve about matters of climate and did not know what brutal winter weather could do to tender plants. I bought vines of a thornless trailing variety in hopes that I could protect them from the winter by laying the supple canes on the ground and covering them with leaves each fall. But my gardening discipline was haphazard at best, and I soon found that Montana winters often arrive early and quickly. My berry crop was so sparse that I was lucky if I could save enough from six vines to make one pie a year. ✑ I have since accepted my casual gardening style and learned to plant hardy raspberries as my sole bramble crop. Raspberries are tough and, at least for a short while, relatively forgiving of careless cultivation. In fact, properly chosen brambles are a perfect crop for busy home gardeners. Brambles produce a crop the second year after 37 planting and can be left in the same place and harvested year after year as long as no serious diseases develop.

One very important reason for growing bramble fruits is that they are almost impossible to find in their fresh state in the grocery store. They are soft and fragile and do not store well, so when they do appear for sale, they are very expensive and often tasteless. Unless there is a pick-your-own farm nearby, the only way to get your fill of these tender delicacies is to grow them yourself. Now that there are thornless varieties, picking their prolific fruits can be an enjoyable task without the bending and stooping entailed in harvesting strawberries. Raspberries will yield about a pint per foot of row over the season, while a single trailing blackberry cane can produce a whopping 15 quarts every summer. A relatively small planting of brambles – a 25-foot row – should provide enough fruit for a family of four.

Members of the genus *Rubus*, all brambles have berries that are actually collections of tiny fruits called drupelets, formed from individual ovaries within the same flower. When the berries are picked, blackberries stick to the receptacle – the part of the flower stalk that holds the fruit – while raspberries separate from the receptacle when ripe so that each berry forms a delicate, hollow cone. Raspberry-blackberry hybrids – such as boysenberries, loganberries, youngberries and tayberries – are tricky to develop, because the blackberry trait of drupelets sticking to the receptacle tends to combine with the raspberry trait of the receptacle sticking to the plant, resulting in fruits that have to be wrestled free even when they are dead ripe.

Raspberries come in four fruit colours: red, yellow, black and purple. In general, red raspberries, which grow on upright canes, either thorny or thornless, are the hardiest of brambles, so some cultivars of these are the most suitable for northern gardens. The few yellow rasp-

berries belong to the same species. Black raspberries, or blackcaps, which belong to a different species, produce smaller, less juicy fruit on vigorous, upright, thorny canes. Purple raspberries, crosses between the red and black species sometimes referred to as separate species, have fruit that is large, juicy and tart on tall, thick canes.

Blackberries come in more variations than do raspberries. The hardiest cultivars have erect, thorny canes, while the more tender trailing types have either thorny or thornless canes. Western trailing blackberries are adapted to the mild climate of the West Coast, while others, sometimes called dewberries, can tolerate the humidity and diseases of the U.S. south. Unfortunately for northerners, the thornless trailing blackberries are the least hardy of brambles. They can be damaged when temperatures dip below 10 degrees F.

Gardeners who understand the way brambles grow have the best chance of success. Each spring, underground buds on the year-old canes send up new shoots called primocanes, which form leaves but not flowers. The shorter days and cooler temperatures of fall stimulate the primocanes to form flower buds, and the next spring, light and warmth urge the flower buds to finish their development by blooming and producing fruit. The second-year blooming canes are appropriately called floricanes. After producing fruit, the floricanes die and should be pruned off at ground level; new primocanes will have grown to replace them. In addition to the canes that come up every year at the base of the plant, other canes grow from root suckers on all types except black raspberries, so the bramble patch spreads out in all directions.

This is the basic growth pattern, but it is modified in some cases. The primocanes of black and purple raspberries and some blackberries produce side branches that bear the flowers. Fall-bearing red and

Most raspberries are ready to pick in early summer, but the fall-bearing types such as

'Heritage' produce a second, late crop or are pruned for only a fall crop.

yellow raspberries, described below, actually produce two crops – one in summer and one in fall.

If possible, choose the site for your brambles a year ahead so that you can prepare it to their liking. To avoid the transfer of diseases, plant brambles at least 200 yards from wild berries. Weeds such as lamb's quarters and red-rooted pigweed can carry verticillium wilt, as can strawberries and crops in the family Solanaceae, such as eggplant, tomatoes and potatoes, so stay away from places where these crops have grown within the past three years. Grass can harbour grubs that eat raspberry roots. Cultivating and weeding the ground the season before planting will help control these problems. If you need to keep the garden in production, corn is a good crop to grow the year before you plant brambles.

Raspberries prefer full sun, but blackberries can take some shade, especially in hot places, where a full day of sun can damage the blossoms and shrivel the berries. Filtered sunlight or shade during the hottest part of the day will keep blackberries from being stressed in such situations. Land that slopes is good for several reasons. Slopes provide good water drainage and air circulation – when the canes are crowded in stagnant air, they are susceptible to a variety of diseases. Cold air flows downward and settles in low spots, so plants on slopes have some protection from late spring frosts that can damage flowers and the slender lateral canes. Brambles planted on a south-facing slope are likely to have ripe fruit earlier than those on a north-facing slope, so where winters are harsh, a north-facing slope is preferable, because warm weather in late winter can awaken the brambles too early and they will be injured if cold weather returns.

Brambles also need well-drained soil 39

Catalogue illustrations make all varieties look tempting, but consider climate first.

rich in organic matter. As their roots grow in the top 20 inches of soil, they may have difficulty getting enough water in sandy soils. But they are also very sensitive to too much water and can die if planted in clay soils that do not drain well. A green-manure crop such as clover or annual ryegrass dug into the ground a few weeks before planting will boost soil fertility and add organic matter.

Allow brambles a weed-free strip of ground 3 or 4 feet wide. If you want paths on either side of the row, allow a width of at least 6 feet. Raspberries should be spaced 2 feet apart, while blackberries need 3 to 4 feet. Black and purple raspberries should be 3 to 4 feet apart as well, because the canes form fruiting lateral branches. Trailing blackberries should be 6 feet apart in the north, where they grow relatively slowly, and 8 to 12 feet apart in the south, where they thrive. If your garden space is limited, you can fill the spaces

between plants with annuals such as peas, beans or cauliflower the first year.

Bramble Choice

Browsing through garden catalogues is delicious torture. Everything looks so good, it is difficult to know what to choose. Unfortunately, many catalogues try to sell all their varieties equally, making choices even harder. But if you take your climate into consideration and know how you want to use your crops, it will be easier to narrow the list of possibilities.

Hardiness is the most important consideration. Brambles bred for cold-winter areas do not thrive in warm ones, and vice versa. The length of the harvest season and the time you want the berries to ripen also dictate choice of variety. Red raspberries, for example, can be harvested from the first week of July (early) through mid-July (midseason) and until the first week of October (fall-bearing). If you live in a northern mountain valley where frost hits in September, you will not want to plant 'Heritage,' whose fall harvest season begins at the end of August and can last into October.

There are dozens of cultivars available – old favourites and new releases – every year. The following list of the more popular and promising brambles notes their strengths and weaknesses.

Red Raspberries
(single-crop, or June-bearing)

'Boyne' was bred in Manitoba and is thus a good choice for cold gardens. It has spiny, 5-foot canes that produce many suckers. The small-to-medium-sized fruit ripens around early July, is aromatic, has fair flavour and freezes well. The canes are susceptible to anthracnose.

'Canby,' from Oregon, is a favourite where I live, in U.S. Department of Agriculture (USDA) climatic zone 4. The vines are almost thornless. The bright red midseason fruit is medium to large, with excellent flavour.

'Dorman Red,' from Mississippi, is a productive cultivar adapted to the south, where it ripens in mid-June on 15-to-20-foot canes.

'Gatineau' is a very hardy Ontario release that ripens early, producing soft fruit with mild flavour. The canes are small, thin and not very vigorous but relatively tolerant of wet soil.

'Latham,' developed in Minnesota, is a popular, widely adapted variety that produces few spines on vigorous canes which are very hardy and virus-resistant. The fruit ripens in midseason and continues for a long harvest. The small, somewhat crumbly berries have fair flavour.

'Willamette' is the traditional Pacific Coast favourite developed in Oregon in 1943, although it is now being phased out by newer cultivars. Tall canes are very susceptible to winter injury. Fruit is large and tasty.

Red Raspberries

(two-crop, or fall-bearing)

'August Red' is the earliest of the primocane-fruiting types, and it has short, spiny canes and mild-tasting, medium-sized fruit.

'Fall Red' produces large, tasty berries on vigorous canes with many suckers. The soft, small-to-medium-sized fruit is good for short growing seasons.

'Heritage' produces a late crop from erect, thorny, 5-to-6-foot primocanes. The fruit is medium-sized, tasty and good for freezing. 'Heritage' is not recommended for areas with cool summers or frost before September 30.

Yellow Raspberries

'Fall Gold' is the only widely available yellow raspberry. It produces very sweet, soft fruit with excellent flavour on vigorous vines with many suckers. The primo-

cane crop ripens relatively early. 'Fall Gold' is susceptible to viruses and too tender for really cold areas.

Black Raspberries

'Black Hawk,' from Iowa, yields large, good-quality berries on thorny, vigorous, upright canes. It is one of the hardiest black raspberries, and it is resistant to anthracnose.

'Bristol,' from New York, produces large, delicious, early-ripening berries on vigorous, productive plants.

'Cumberland,' from Pennsylvania, is susceptible to fungal and viral diseases and not very winter-hardy. The midsea-

Unlike blackberries, black raspberries separate from the receptacle when picked.

son fruit is a moderate size with fair flavour on vigorous canes.

Purple Raspberries

'Brandywine,' from New York, yields large, late, tart fruit on vigorous, thorny, 8-to-10-foot canes that are nonsuckering and somewhat disease-resistant.

'Royalty,' also from New York, considered the best purple berry by many growers, produces large, late, sweet berries on vigorous, thorny canes that are 41

Blackberries come in more variations than do raspberries. The hardiest have thorny, erect canes, while the trailing types may be either thorny or thornless.

resistant to aphids and viral diseases.

'Sodus' yields large, tart, midseason berries on productive canes.

Blackberries

(erect)

'Brazos' crops early, continues for a long time and is somewhat disease-resistant. It thrives farther south than most other erect varieties, so it is popular in Louisiana, Texas and Arkansas.

'Darrow' produces a heavy crop of very large, fine-quality midseason berries over a wide geographical area.

'Ebony King' yields early, purplish, tangy fruit on canes that are relatively hardy and resistant to orange rust.

'Rosborough' is similar to 'Brazos,' but its fruit is firmer and has better flavour. It produces an early, heavy crop in warm-winter areas.

Blackberries

(trailing)

'Black Satin' produces a midseason crop of large berries on thornless, disease-resistant vines. It grows in a wide variety of climatic areas.

'Chester' is a new midseason thornless variety that will grow in cooler areas than most trailing berries – around the southern Great Lakes, for instance. The berries are firmer and less tart than those of 'Black Satin.'

'Hull' is relatively hardy for a thornless trailing variety and will grow in gardens around the southern Great Lakes and inland Pacific areas. The midseason crop of sweet berries is abundant.

'Lucretia' is an old favourite that does well in the south. The berries are early and extra large.

'Thornfree' produces an abundant crop of large, tangy berries on disease-resistant vines late in the season.

Raspberry-Blackberry Hybrids

Boysenberry, whose origin is unknown, is a thorny Pacific Coast berry that can also be grown in the south. The tangy, purplish fruit is produced earlier than that of the midseason youngberry. There is also a thornless type.

Loganberry, the result of an unplanned cross that sprouted from blackberry seedlings planted by Judge Logan in California in the 1890s, is also available in both thorny and thornless versions. This is a good West Coast berry whose late, reddish, tart fruits are delicious in pies and preserves.

Tayberry, a recent Scottish cross that produces tangy, dark red berries in midseason, is somewhat hardier than the former two berries and will grow in warmer gardens around the southern Great Lakes and inland coastal regions. The fruit is larger, earlier-ripening and less acidic than a loganberry.

Youngberry produces large, early, purplish black midseason fruits that are much sweeter than loganberries. It is good in very mild gardens in the south and on the Pacific Coast. A thornless version is available.

Planting

Brambles are usually planted in spring. If you buy them by mail, they will arrive looking quite dried out and dead, like short sticks attached to tangled masses of roots. To give the plants a head start, soak the roots in cool water for a couple of hours before planting. Bring the bucket into the garden so that the plants can be moved directly from the water into the ground. For each plant, dig a hole bigger than the root mass, and spread the roots out as you position the plant in the ground. Cover the roots with soil to the same level as the plant grew before, which is indicated by a brown line on the lower part of the cane. Water the plants well, and make sure they continue to receive plenty of water for the next week or two while they become established. Mulching, which keeps weeds down, is a key to success with brambles. Apply 3 to 8 inches of organic material such as straw, shredded leaves or sawdust mixed with manure.

Trellising

Setting up a trellis for brambles requires planning, elbow grease and elementary carpentry skills, but it pays off in the long run. A good, sturdy trellis will keep plants under control and increase the yield of berries considerably.

Red raspberries can be supported in several ways. The traditional trellis is T-shaped, with horizontal wires strung along the outsides of the crossbars to keep the canes from falling down when they are heavy with fruit. But a newer method, a V-shaped trellis, can increase yields enormously and make harvesting easier. Create a V trellis by adding a lower cross arm to a T trellis and stringing another set of wires (see diagram) or by pounding 7- or 8-foot metal fenceposts into the ground every 25 to 30 feet along the row. The metal hooks should be on the outside of posts set at an angle of about 30 degrees,

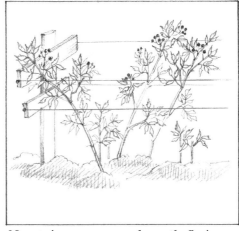

New primocanes grow beneath floricanes trained against the wires of a V trellis.

18 inches apart at the base and 3½ feet at the top. Using the metal hooks on the posts to hold the wire, string the trellis with heavy galvanized wire or 2½-millimetre monofilament line, one strand about 24 inches from the ground and the other about 40 inches. To stabilize the trellis, set additional posts a couple of feet beyond the end posts to anchor the wires.

In early spring, tie the second-year canes to the trellis wires to keep them on the outside. Pull the floricanes to either side to form the sides of the V, and let the 43 primocanes grow up through the centre so that they will receive plenty of sunlight and air without being disturbed during the harvest. An alternative that works with the metal-fencepost trellis is to run an additional wire down the middle of the row of canes and use it to push half the canes to one side. Anchor this wire to the posts halfway between the wires that are already in place. Next, string a second wire to hold the remaining canes to the other side of the trellis.

You can support trailing blackberries in several ways. Simply driving a sturdy pole next to each plant and tying the long canes to it will work, but a trellis gives the vines more light and air and makes picking easier. Posts should be 15 to 20 feet apart, with horizontal wires strung 3 and 5 feet high. In mild-climate areas, tie the primocanes to the wires when they are sufficiently long. In areas with more severe climates, let the canes grow along the ground and tie them to the wires in spring, fanning out the canes. Long canes can be looped over the top wire and spread out along the bottom one. A V trellis can also be used for trailing blackberries, in which case, the fruiting canes are tied to one side of the V, the primocanes to the other.

Pruning

After the harvest, the old fruiting canes of all brambles (except perhaps fall-bearing raspberries, described below) should be cut off at ground level and burned or otherwise disposed of away from the planting. This gives the primocanes more space and light and gets rid of any diseases and pests in the old canes.

Beyond this simple rule, bramble-pruning techniques depend on type. Some books recommend pruning single-crop red and yellow raspberries back to 4 feet tall in fall. If you do not trellis your raspberries, cutting them back in this fashion will help keep them from falling over when laden with fruit, but such pruning removes the most productive part of the cane and drastically reduces yield. The berries will be bigger, but the total volume of the crop will be much smaller. Under optimal conditions, raspberries can produce canes more than 6 feet tall. Since 6 feet is about as high as most adults can comfortably pick and since the most fruitful buds develop along the middle three-fifths of the cane, this is a good height to aim for when pruning raspberries.

Fall-bearing red and yellow raspberries are treated differently. The tops of the primocanes should be pruned off after they bear the fall crop. The next summer, these canes will produce berries just like the floricanes of other raspberries. If you live in a mild area, you might want to grow fall-bearing raspberries for their autumn crop alone. In this case, wait until after the first heavy fall frost, when the stored nutrients in the canes have retreated into the roots, to cut all the canes down to the ground. New primocanes will grow in spring and bear in fall. You will harvest fewer berries overall this way but will pick a larger fall crop (about one-half to three-quarters of a quart from each plant) while avoiding most pests and diseases. Because there is only one pruning – when all the canes are cut down – the method is simple and streamlined. In fact, it is becoming so popular with commercial growers that new varieties are being created specifically for it.

Black and purple raspberries produce flowers on side branches that grow from primocanes the first year. If you let the primocanes grow, a hormone produced by the growing tip will suppress growth of the lateral branches, reducing the yield. But if you prune the tips off the primocanes when they are 18 to 24 inches tall, they can produce lateral branches several feet long. Prune the laterals to about a foot long in late winter to make them produce abundant fruiting branches. A few black raspberries flower mainly near the tips of the laterals; do not prune them so severely.

Black raspberries produce their fruit on side branches that grow from the primocanes.

Upright blackberries also fruit on laterals. Their primocanes should be tipped when they reach 3 to 4 feet high, and the laterals can be cut to 12 to 16 inches in late winter or early spring, as with black raspberries. While cutting out the old canes, reduce the number of canes per plant to improve air circulation and let light into the centre of the plant. Prune away the thinner canes, leaving the five or six strongest per plant, since the thickest produce the most fruit. Watch for suckers growing far from your rows; these should be pulled out because they will regrow if they are only cut back.

Trailing blackberries and dewberries should also be thinned after the harvest to about eight canes per plant. How you prune these berries depends on how old the vines are and where you live. For the first two years after planting, the canes of trailing blackberries are flexible and grow along the ground if not supported on a

trellis, but by the third year, the plants produce stiffer canes that will not bend to the ground, so you can no longer protect them in winter. If you live where winters are cold, tip the primocanes when they reach 24 inches long. This will encourage the development of laterals close to the ground, where they can be covered for winter. In spring, the fruiting laterals can be brought up to the trellis and woven around or tied to the wires. Remove the laterals closest to the ground.

Care and Maintenance

While brambles are basically easy-care crops, they do need some ongoing attention, including consistent watering, especially during flowering, ripening and fall flower-bud development. Overhead watering, however, can promote diseases and weaken berry flavour, because the berries will soak up water that lands on them. Better is drip irrigation, a system of perforated pipes that provides a reliable source of moisture without waterlogging the plants. If you water by hand, point the hose toward the soil around the plants.

Brambles also need mild fertilizing to continue to yield well. Each spring before the plants begin to grow, pull aside the mulch and spread a generous layer of manure along each side of the row. Do not work the manure into the ground, because brambles have shallow roots. Water and earthworms will carry the nutrients down into the soil. Avoid poultry manure, which can be salty; brambles are sensitive to salt. Every few years, add a pound each of rock phosphate and greensand or bone meal for every foot of row to provide potassium and phosphorus.

In the most northerly gardens, all domestic cane fruits should be protected over the winter unless you know from experience that they will survive. Bend the canes as close to the ground as possible, and cover the tips with enough soil to hold them in place. Begin uncovering before spring growth starts, when minimum temperatures have risen to about 5 degrees F. Where snowfall damage is a danger, tie bunches of canes together, tepee-style, and snip the twine as soon as the snow melts in spring.

Brambles proliferate, so be sure they do not become overcrowded as the years go by. The guidelines for spacing the canes vary depending on the crop. Red and yellow raspberries should be kept to about 6 canes per foot of wide row. Since black and purple raspberries do not form suckers, they will stay where planted, but each year, more canes will grow. In rich soil, where the plants bear well, leave 10 canes per plant. In poorer soil, you will get a better yield if you limit the plants to 5 or 6 canes. Erect blackberries will fill in like red raspberries; allow 5 or 6 canes to every foot of wide row. Trailing blackberries should not be allowed to become crowded, so remove the suckers. Each hill should have no more than about 14 canes.

Plants that sucker are the easiest to propagate – just remove the suckers while the plants are dormant, digging deeply enough to get all the roots. To avoid injuring the mother plants, do not take any suckers closer than about 6 inches. Black and purple raspberries can be propagated easily by tip layering. Bury the tips of the primocanes 3 inches deep in late summer or early fall. Roots will form underground, and small plants will appear around the buried cane. Then, while the plants are still dormant in late winter, cut the mother cane, leaving about 8 inches attached to the new roots. Dig up the plant, taking a ball of soil around the roots. After planting, remove the old cane at soil level.

Problems

When purchasing new plants, buy disease-free stock if at all possible. Brambles are generally trouble-free in the home garden, but they do suffer from some diseases that have no cure. If you always cut

Blackberries are best-tasting when they turn from shiny to slightly dull.

down old canes after harvest and keep the patch well thinned, pests and diseases can be kept to a minimum. If you have any doubt about the health of your plants, do not multiply them, as they will just take their problems with them. Watch for spotted, wilted or yellow leaves, purple streaks on leaves or canes or white mould growing on the plants. If you see none of these telltale signs of trouble and the plants have dark green leaves and are producing well, they are healthy.

Plants infected by raspberry mosaic have greenish yellow leaves or mottled green leaves on stunted canes. The leaves are usually smaller than normal and may be deformed, and the tips of young shoots may turn black. Each year, the canes become more stunted until they die. Black raspberries are especially susceptible, while red raspberries are not as seriously affected. Blackberries are not generally victims of this deadly disease, which is spread by the large raspberry aphid. To avoid raspberry mosaic, plant black raspberries far from old plantings of raspberries and from wild plants. Infected canes should be removed and burned immediately. Some newer red and purple varieties are resistant to the aphid and therefore unlikely to suffer from mosaic.

Three different diseases can cause crumbly red or yellow raspberries, but one of these is rare. If the berries crumble, it may be simply a question of proper care; poor pollination, nutrient deficiencies or water stress can cause the problem. Unfortunately, one of the viruses responsible for the symptom, raspberry bushy dwarf virus, has taken over much of the stock of the only widely available yellow raspberry, 'Fall Gold.' If you want to grow this cultivar, be sure to buy certified virus-free stock.

Leaf curl, another virus, reduces leaf size while causing the edges of the leaves of black and red raspberries to curl downward. Infected canes will not grow well and will produce small, crumbly fruit. Purple raspberries are less susceptible, and most blackberries do not show symptoms of this disease, which is carried by the small raspberry aphid. Control the same way as for mosaic virus.

Orange rust is a common disease of blackberries, causing bright orange lesions on the canes and undersides of the leaves in summer. Wild vines often carry the disease, and nurseries may inadvertently sell infected plants, so beware. Immediately destroy any stock that shows symptoms; there is no cure.

Black raspberries are especially vulnerable to verticillium wilt, another fungal disease that has no cure. Leaves turn yellow-bronze, then die or fall off. Canes may show purple or blue streaks and may wilt and die before a crop matures. Red raspberries and blackberries have some resistance to this disease but can also be affected. Nursery stock can carry verticillium wilt even if it looks healthy. Do not plant brambles where strawberries or members of the tomato family have recently grown.

Black raspberries are also most susceptible to anthracnose, a fungal blight that causes sunken purple or greyish spots on canes or small, grey leaf spots with purple margins. The centres may fall out,
leaving holes in the leaves. Anthracnose is worst on new growth. Some black raspberry varieties are resistant.

The cane borer, the only common serious insect pest, is at work if the tops of new canes suddenly wilt. Close examination will reveal a circle girdling the cane near the top. Just beneath this circle lies the borer, which is the larva of a beetle. If you can catch the borer at this early stage, eliminating it is simple: just cut off the cane about half an inch below the girdle. If you do not remove the grub, it will burrow down through the cane, feeding as it goes, passing its first winter just below the girdle its mother made. By the second winter, it will be near the ground.

Harvest

Picking bramble fruits is a pleasure, especially if the plants are thornless. The berries ripen quickly on the canes, so check the bramble patch every other day to avoid wasting fruit. Raspberries are ripe when they come off the plants easily, while blackberries are at their sweetest and juiciest just after they turn from shiny to slightly dull. Pick the berries in the morning, when their sugar content is highest, and if possible, avoid picking directly after rain, as water soaked up by the fruit dilutes its flavour.

Raspberries and blackberries do not store well, so eat them, cook them or freeze them within a day of picking. Freezing these fruits is easy. Spread them on cookie sheets, and place them in the freezer. When the fruits are solid, transfer them to sturdy plastic bags. The fruits will remain separate, and you can measure out the desired quantity with ease.

Dorothy Hinshaw Patent, who has been an enthusiastic gardener since moving to Montana in 1972, is co-author of *Backyard Fruits and Berries* (Rodale Press) and *Garden Secrets*, originally published by Rodale and soon to be reissued by Camden House. 47

Chapter Four:
Wild Berries

By Nancy J. Turner

ild berries. The very words evoke vivid images of foraging for fragrant, juicy strawberries, collecting enough huckleberries for a tangy pie or harvesting buckets of saskatoons, elderberries or blackberries for jelly, juice or wine. Although wild berries are usually associated with hiking, camping or country living, many types grow in vacant city lots or on suburban fringes, and a number will thrive in home gardens. In the interests of species conservation, of course, wild berries should not be transplanted from their native habitat, but most can be propagated easily from cuttings, and many grow well from seed. Some are available from retail nurseries. ✑ There are more than 200 species of small fleshy fruits – called berries by most people – growing wild in Canada and the northern United States. Technically, only a few of these are rightly called berries. Fruits such as highbush cranberries and bunchberries, which have a single seed or stone surrounded by fleshy tissue, are known as drupes. (Cherries, peaches and plums are also drupes.) Saskatoon berries are classified as pomes, like apples and pears. Raspberries, blackberries and strawberries are known as aggregate fruits. Finally, there are the true berries, including blueberries, currants and wild grapes, which have many small seeds scattered within soft, fleshy tissue and are usually globular in shape. In this 49 chapter, I use the term "berry" as it is used throughout this book, in its less technical sense, to refer to any small fleshy fruit. I have not included wild cherries and plums (*Prunus* spp), however, because they are not generally considered berries.

One of my greatest interests is the use of berries by indigenous people. Throughout the continent, even in the far north, wild fruits have been used since antiquity in traditional native diets. Most berries are rich in vitamin C, but they also contain carbohydrates, proteins, calcium, iron, vitamin A and thiamine, as well as riboflavin and niacin equivalents and trace elements. The presence of these nutrients and the fact that some berries remain attached to the bushes over winter make certain wild berries important emergency foods. For people stranded in the wilderness, a good knowledge of edible wild berries can mean survival.

Wild berry crops vary in quality and productivity from one place to another, and some produce generously in some years and very sparsely in others. Wild strawberries are often among the first fruits to ripen. Saskatoons are also relatively early, followed by some of the wild blueberries and huckleberries, raspberries, blackberries, gooseberries, currants and bunchberries. Finally, in late summer and fall, come salal berries, blue elderberries and various types of cranberry, along with rose hips and some later-fruiting species of huckleberry, gooseberry and currant. Many of the more tart fruits, such as silver buffalo-berry and highbush cranberry, become sweeter after the first fall frosts.

Most berries can be preserved for year-round use. Drying or dehydrating is a common traditional method. Saskatoons, blueberries and other fruits were sometimes dried on mats, like raisins. Sometimes, too, berries were mashed and partially cooked, and the jamlike pulp was spread in the sun or over a fire to dry in cakes or loaves. Dried berries mixed with dried salmon or meat and fat made pemmican, a nutritious, high-energy, high-protein food. Nowadays, the drying process is faster and easier with a solar or electric food dehydrator, but wild berries are usually preserved as jams or jellies or by canning or freezing.

There are no infallible rules for distinguishing poisonous berries from edible ones. With the help of a knowledgeable teacher or a good reference book, however, an observant person can quickly learn to recognize the common edible and inedible berries of a region. Never eat a berry (or other plant) unless you are sure of its identity and edibility and any special preparation required. Even the best edible berries can be rendered potentially dangerous if they are gathered from areas polluted by pesticides, vehicle exhaust or other emissions. Before picking berries, find out if pesticides have been sprayed in the area. Even if the berries along a busy highway look delicious, they may be contaminated with lead or other poisonous compounds. Finally, remember to respect the rights of property owners when picking berries, and be sure to leave behind enough for the birds and other wildlife.

Amelanchier alnifolia

(saskatoon, serviceberry)

The Cree name for these berries, approximately rendered *mis-sask-qua-too-min*, gave the city of Saskatoon its name — a compliment to both the city and the berries. There are about 15 species of *Amelanchier* in Canada, all but the species *alnifolia* being eastern in distribution. They are variously called Junebush, shadbush, Chuckley pear, sugar pear, Indian pear and grape-pear. All have sweet, juicy fruits that, as the last four names imply, somewhat resemble pears in flavour if not appearance.

Saskatoon berry, which is known as serviceberry on the U.S. side of the border, is one of the most popular fruits of the

commercial berry production in Alberta and Saskatchewan.

Arctostaphylos uva-ursi

(kinnikinnick, bearberry)

This attractive evergreen shrub is becoming increasingly popular as a spreading ground cover for landscaping. It thrives in gravelly or sandy, well-drained soil in sun. Kinnikinnick has dark green, leathery leaves that are oval and smooth-edged. The pale pink flowers, like many of those in the heather family, are urn-shaped and hang in small clusters from the ends of the twigs. The fruits are bright red

prairies. The bush ranges from knee-high to much taller than a person, and it has multiple stems, reddish grey bark and round to oval leaves that are generally toothed around the top half and smooth around the base. The white, five-petalled flowers are borne in showy clusters, usually in May and June. The fleshy purplish to dark blue fruits, which resemble large blueberries with sweet, juicy pulp, can be eaten raw or cooked in muffins, pies or sauces. The berries preserve well and are notably rich in iron.

The saskatoon is becoming recognized as a valuable landscaping plant; English gardeners have known about it for many years. Not only is the shrub attractive in appearance, but it is an excellent source of food and shelter for wildlife and is tolerant of many different growing conditions. It can be propagated from seeds, although these require cold stratification for a period of from three to six months. Cuttings should be treated under mist for best success. Fortunately, many nurseries now stock saskatoons for home gardens. Several varieties have been developed for

outside, with a whitish, mealy inner pulp surrounding several large nutlets fused together as a single stone. The name kinnikinnick is said to be derived from an Algonquian word for smoking mixture, because other than tobacco, this was probably the plant most widely used for smoking among North American indigenous people. Kinnikinnick grows on open, rocky slopes and sandy areas throughout Canada, north to Alaska, south to California, west to New Mexico and east to Virginia and New York. Related species, also 51 with edible berries, grow from the Arctic to California.

The berries can be harvested in summer but remain on the plants throughout the winter and even into spring, making them an important famine and emergency food. Because the berries are dry and mealy, they are not considered a prime edible type. Nevertheless, they were widely eaten by indigenous people, usually cooked in fat or oil or boiled in soup to make them less dry. They are relished by deer, grouse and bear.

Kinnikinnick can be grown readily from cuttings or layered shoots treated with a rooting compound. Growing the plant from seed is more difficult, because the seeds are said to germinate only after treatment with sulphuric acid.

Berberis aquifolium

(Oregon-grape)

Also known by the botanical name *Mahonia aquifolium*, Oregon-grape is an evergreen shrub whose common name pays tribute to the plant's abundant growth in Oregon. If it had been named for its leaves, it would have been called Oregon-holly, because the plant resem-

bles holly, with its shiny, leathery, prickly edged foliage. But unlike holly leaves, those of Oregon-grape are compound, each consisting of a central stalk with four to eight leaflets and a single terminal one. Underneath the light grey bark is a layer of brilliant yellow caused by a compound called berberine.

The small flowers, produced in spring in golden yellow masses, give way to dark blue, pea-sized berries in clusters like small grapes. The berries, which ripen in midsummer, are juicy but very tart. With the addition of sugar or other sweeteners, they yield a delicious jelly and can also be used for wine. One of the best jellies I ever made was from 2 cups of Oregon-grapes and 2 cups of salal berry juice, mixed with 4 cups of sugar and a box of pectin crystals, according to general instructions for making cooked jelly.

Oregon-grape is native from southern and central British Columbia and southwestern Alberta to Oregon and Idaho, where it grows in well-drained soil in open woods and clearings and on rocky outcrops. In Britain and other places beyond its native range, it is valued as an ornamental. Oregon-grape is easy to propagate and fast-growing, requires little care and has attractive blossoms, berries and year-round foliage. Because it is prickly, it serves well as a screening or hedge plant 3 to 6 feet tall. Low or creeping varieties are also available. Oregon-grape is best grown from seeds, which germinate well after cold stratification, but it can also be propagated from cuttings, especially if they are taken in the fall.

Cornus canadensis

(bunchberry)

This small herbaceous perennial, also known as dwarf dogwood or crackerberry, grows in woods and damp clearings across Canada and the United States and as far south as New Mexico. It seldom grows more than 8 inches tall but has

many branching rootstocks, so it often forms dense patches. The leaves are oval and pointed, the upper ones forming a whorl at the top of the stem. The flowering heads, which consist of four white bracts surrounding a greenish cluster of tiny flowerlets, are miniature versions of those of flowering dogwood (*C. nuttallii*), a tree-sized relative that bears the provincial flower of British Columbia. A closely related species, *C. suecica*, is very similar but has dark purple flowerlets and smaller bracts. The globular fruits of bunchberry are tightly clustered, orange to bright red and very attractive. While seldom eaten today, the fruits are pleasant-tasting although a bit pulpy, with a hard central stone. Nevertheless, native people ate them raw and fresh, simply discarding or swallowing the seeds.

On the Pacific Coast, bunchberry often grows on logs and trunks of western red cedar. According to one myth, bunchberries originated from the blood of a young woman who had been stranded at the top of a cedar tree by her jealous husband. When her brother, who had come to res-cue her, saw the berries arising from the blood that had run down the tree, she told him, "Do not be afraid. You can eat this. I made it. It is good to eat. It will always grow around the cedar tree."

Bunchberry makes an excellent ground cover in a moist, acidic, somewhat shady garden. It requires a rich, humusy soil, with rotten wood and a little sand mixed in. It can be grown from sections of creeping rootstock but is best started from seed in peat pots, then transplanted out once the seedlings are well established.

Empetrum nigrum

(black crowberry)

"Easily the most important fruit of the Arctic," according to Canadian botanist A.E. Porsild, black crowberry is so hardy that it is well adapted to the windswept slopes and hills of the northern tundra. It bears juicy, mild-flavoured berries, sometimes called blackberries by Inuit and other northern people. These berries are generally deep purplish black but may be pink, bright red or reddish purple. They 53 can be eaten fresh or stored by freezing, which is what the Inuit do, or they can be used in pies and jellies. Porsild has found that a sparkling white wine can be produced by fermenting the juice.

Black crowberry is a low, mat-forming shrub whose short branchlets and dense, spreading, needlelike leaves resemble those of a coniferous tree. The plant, which grows in extensive patches in peat bogs and on rocky mountaintops from the Arctic Ocean through Alaska and Canada to California and New York, is easily mistaken for heather, although it is unrelated. The flowers are small and inconspicuous.

Crowberry makes an excellent ground cover in an alpine garden and is easily propagated from cuttings. The edible berries add to its attractiveness.

Fragaria spp

(wild strawberry)

Wild strawberries look so much like cultivated ones that few people would fail to recognize them. There are the usual long-stalked, three-part leaves, coarsely toothed along the margins, the white, five-petalled spring flowers and, of course, the scarlet, juicy berries, which are much smaller than those of domestic cultivars. Wild strawberries are no less fragrant or flavourful, however, and in fact are often considered tastier.

Two of the most common species, *F. vesca* and *F. virginiana*, are widespread in Canada in open woods, meadows and clearings from sea level to mountain slopes. The two species can usually be distinguished by leaf colour: *F. vesca*, known as common, or woodland, strawberry, has bright green or yellowish green leaves, whereas the leaves of *F. virginiana*, the Virginia, or field, strawberry, are bluish green. Both species produce long runners, or stolons, enabling them to form extensive patches. A third species, *F. chiloensis*, has thicker, shinier leaves and grows on sand dunes and rocky headlands along the Pacific Coast. All three species, especially the last, have been used by plant breeders to develop cultivated varieties.

Wild strawberries can be made into jam or baked in pies like any other fruit, but they are so soft that, when fresh, they do not transport or store well. Indigenous people often dried them for winter storage and used them as a sweetener for other berries. They freeze well.

Wild strawberries are easy to cultivate and are excellent ground-cover plants, spreading rapidly by their creeping runners. The best way to establish them is to transplant rooted offshoots from the ends of the runners. The plants, which thrive in mowed meadows and clearings, require little ongoing care.

Gaultheria shallon

(salal)

David Douglas, the Scottish botanist for whom the Douglas fir is named, first encountered salal in 1825 when, after a tedious sea voyage of 8½ months, he rowed ashore by the mouth of the Columbia River. There, he found this attractive evergreen shrub: "So pleased was I that I could scarcely see anything but it. . . . It grows under thick [conifer] forests in great luxuriance and would make a valuable addition to our gardens." Salal, with its tough, resilient, branching stems, is an important understorey plant in West Coast forests. It varies from knee-high to waist-high and, in some places, is much taller, occasionally forming patches so dense as to be virtually impenetrable. The oval, leathery, shiny leaves are pointed and finely toothed around the margins. The urn-shaped, light pink to whitish flowers are borne in long, one-sided clusters, with hairy, somewhat sticky stems. The berries are dark purple to nearly black, juicy and sweet, with many tiny seeds. At the tip of each berry is a starlike depression.

For indigenous people of the Pacific Coast, from southern Alaska to Washington, salal berries have been the most important and widely used of the wild fruits. Some were eaten fresh, but most were stored for winter. Today, many people, both native and nonnative, use salal ber-ries for jams, jellies, juices and preserves.

There are several other small *Gaultheria* species, all with sweet, edible berries. Wintergreen (*G. procumbens*) is an attractive, mat-forming evergreen with bright red wintergreen-flavoured berries that remain on the plant throughout the winter. The berries are edible, and a tea can be made from the leaves, but because wintergreen oil is chemically related to acetyl-salicylic acid (the active substance in aspirin), wintergreen should be avoided by people allergic to aspirin.

Douglas's suggestion about the value of salal in landscaping is finally becoming reality, especially for gardens in forest clearings. In the Pacific Coast region, many nurseries now sell salal. The seeds germinate well in a mixture of moist peat and sand, but it takes several years for the plants to grow large enough to set out. A slightly faster method of propagation is to take cuttings of new wood in late summer, coat them with rooting compound and start them in moist sand.

Morus rubra

(red mulberry)

Most of us are better acquainted with the children's song "Here We Go Round the Mulberry Bush" than with the mulberry itself. In Canada, red mulberry, a bushy, deciduous tree, is native only to the forest region of southernmost Ontario, near the shores of Lake Ontario and Lake Erie, although it has been planted beyond this region as an ornamental and has escaped from cultivation in parts of southern Ontario and British Columbia. It is more widespread in the United States, where its range extends from the Great Lakes south to Texas and Florida.

Growing on slender stalks, mulberry leaves are large—often longer than 5 inches—toothed along the margins and often variously lobed. The male and female flowers, greenish and inconspicuous, are usually borne in separate clusters, before 55

shelter for wild birds. It may be grown from seed, from cuttings or by layering and is sold by some nurseries. It is not particularly hardy, but it can be grown beyond its natural range. Consult local nurseries to find out whether it will grow in your area.

Oxycoccus spp

(cranberry)

or with the leaves in spring. The fruits, dark reddish purple to almost black, resemble blackberries. When ripe, they are sweet and juicy, but the unripe berries and the milky sap in the leaves and stems are toxic and may cause stomach upset. The sap also may irritate the skin.

A closely related Asian species, white mulberry (*M. alba*), has spread from cultivation in parts of eastern North America, including southern Ontario. In one form, the berries are whitish to pinkish; in another, they are dark purple to black. Its leaves are shiny and generally smooth underneath, unlike those of red mulberry, which have rough, hairy undersides. In eastern Asia, white mulberry is the main food of silk-moth caterpillars. It was first introduced to North America more than 200 years ago in an attempt to begin a silk industry.

Mulberries can be eaten fresh or made into juice, jelly or jam. They are excellent in pies, muffins and pancakes and mix well with other fruits in baking or jellies. The mulberry is also a valuable shade and ornamental tree, providing both food and

Several types of wild berries are known as cranberries. These include bog cranberry (*Oxycoccus* spp, close relatives and precursors of the cultivated cranberry), lingonberry (*Vaccinium vitis-idaea*, sometimes called mountain, lowbush or rock cranberry – see page 64) and highbush cranberry (*Viburnum edule* – see page 65). Bog cranberry and lingonberry, members of the heather family, are closely related to blueberries and huckleberries. Highbush cranberry, which belongs to the honeysuckle family, is similar in flavour but very different in appearance.

Four species of bog cranberry are recognized in Canada and the northern United States. They are difficult to distinguish, so I will describe them together. All are low, slender, creeping, evergreen vines that grow in acidic peat bogs and muskegs usually associated with sphagnum moss. Their branches are thin and flexible and their leaves tiny, oval to elliptical, smooth-edged and tending to curl under at the edges. Their flowers, borne on threadlike stalks, are small, pink, nodding and four-parted, with recurved petals and prominent stems. The berries are elongated or globular, purplish to bright red when ripe, at first firm and acidic but softer after the first frost. The largest-fruited species, *O. macrocarpos*, is the main forebear of the cultivated cranberry.

Few gardeners would find it practical to grow bog cranberries, since they require such specialized conditions. Commercial cranberries are undeniably more convenient. Nevertheless, the delicate and interesting wild species would be suitable for a bog garden, especially in northern Canada. They can be propagated easily from seeds, cuttings or root offshoots.

Physalis spp

(ground cherry)

Ground cherry, also called husk tomato, is related to the true tomato, and while its ripe fruits are cherrylike in size, they taste somewhat like tomatoes (see Chapter One). The most common native plant, *P. heterophylla*, is erect and herbaceous and grows to 3 feet tall in the rich soil of the prairies, pastures and open woods of southeastern Canada from Manitoba to Nova Scotia and south to Texas and Georgia. The leaves, as long as 4 inches, are oval and hairy on both surfaces, with wavy or toothed edges. The flowers are solitary, with yellowish petals fused into a bell-shaped tube. The ripe fruits are fleshy, yellowish berries encased within an inflated, papery, lanternlike husk. Unripe ground cherries are bitter and possibly poisonous, but when fully ripe, usually in September and October, they are very sweet and agreeable. Ground cherries were widely harvested by native people.

There are at least three other introduced species – *P. alkekengi*, *P. ixocarpa* and *P. pubescens* – and one species native to Canada, *P. virginiana*. The leaves, roots and unripe fruits of all of these should be avoided, but the ripe fruits are edible.

Ribes spp

(wild currant)

About 15 species of wild currant grow in Canada and the northern United States, some with red berries (notably the northern red currant, *R. triste*), others with blue or black. All have edible fruits, as do their relatives the gooseberries. Northern red currant, which is widespread in northern Canada and Alaska along streams and in moist woods, is one of the most popular. It is similar in appearance and flavour to the cultivated red currant, which was derived from a closely related European species, *R. sativum*. *R. triste*, too, grows about 4 feet tall with spreading branches. The purplish flowers in elongated clusters are small but attractive. The shiny, red, jewel-like fruits ripen in August and can be 57 eaten fresh, used in baking or made into delicious jams and jellies.

Currants vary widely in appearance, but all are woody shrubs with palmately lobed leaves. Unlike gooseberries, the shrubs usually lack spines or prickles (an exception is the black-fruited swamp or bristly currant, *R. lacustre*), and their smaller berries are usually borne in bigger clusters.

Of the black-berried species, northern black currant (*R. hudsonianum*) and American black currant (*R. americanum*) are notable for their close resemblance to the black garden currant (*R. nigrum*). Two currant species with strikingly beautiful flowers grown as ornamentals well beyond their natural range are red flowering currant (*R. sanguineum*) and golden currant (*R. aureum*). These shrubs are even more appealing because their showy flowers attract hummingbirds and their edible berries are sought by other birds.

Most currants can be grown in the garden, and they propagate well from layered shoots and cuttings. They can also be grown from seed, which may require two or three months of cold stratification. Some currants and gooseberries serve as alternate hosts to white pine blister rust

fungus: only disease-free stock should be introduced into a garden.

Ribes oxycanthoides

<small>(wild gooseberry)</small>

Cousins of the currants described previously, gooseberries are distinguished by spiny or thorny stems that make the berries difficult to harvest. (Some botanists prefer to use the genus name *Grossularia* for gooseberries, applying *Ribes* only to currants.) About a dozen species of wild gooseberry grow in Canada and the northern United States. Of these, *R. oxycanthoides*, the Canada, or smooth, gooseberry, is probably the most widespread. It is a low, deciduous shrub that grows on rocky or sandy ground along shores, on slopes and in open woods from eastern British Columbia to the Maritimes and Newfoundland and south to Montana, Ohio and Pennsylvania. The small, greenish yellow, bell-shaped flowers are borne in groups of two or three, and the berries are globular, smooth and bluish black, each with a characteristic "wick" at the

end – the remains of the flower parts. Other gooseberry species may have reddish, purplish or black berries with smooth, hairy or bristly skins.

All gooseberries are edible. They make wonderful jams, jellies, preserves, tart fillings and cheesecake toppings. Indigenous people sometimes picked the berries green and pickled them to make a sauce that was eaten with other foods.

In the past, most commercial gooseberries were derived from a European species (*R. grossularia*), but some hybrids and cultivars have more recently been developed from American species, including the coastal black gooseberry (*R. divaricatum*) and currant gooseberry (*R. hirtellum*). These strains have a natural resistance to American gooseberry mildew, a disease that has plagued the European cultivars. Any of the wild species can be grown in a garden setting, both as ornamentals and for their fruit. They can be propagated from seed or from cuttings or layered shoots. Most thrive in gravelly, well-drained soil.

Rosa spp

(wild rose)

There are more than a dozen species of wild rose in Canada and the adjacent states. All are prickly deciduous shrubs with leaves similar to the cultivated types, showy whitish to pink flowers that are normally five-petalled and orange to deep red berrylike fruits called hips. *Rosa acicularis*, prickly rose, one of the most widespread species and the floral emblem of Alberta, grows in open woods and thickets and on rocky slopes across northern Canada from the interior of British Columbia across the Prairie Provinces to Quebec, extending north and west to the Yukon and Alaska and south to New Mexico, Minnesota and Vermont.

Rose hips consist of an edible, fleshy but firm rind enclosing a tight cluster of whitish seeds called achenes. In most species, the seeds are covered with many sharp, sliverlike hairs, which should not be eaten since they can irritate the skin and digestive tract. Either scoop the bristly seeds from the halved hips, or cook the hips and then strain out the seeds for rose hip tea, jelly or syrup. To make tea, place a generous handful of the dried hips, whole or crushed, in a teapot, cover with boiling water and allow to steep for about 10 minutes. For jelly, rose hip juice mixes well with that of crab apples or wild berries such as salal and blueberries. Also, a pleasant, slightly resinous tea can be made from the young twigs and leaves.

Rose hips, well known for their high vitamin C content, are an important food for wildlife and were regarded as emergency or famine food by indigenous people because the hips remain edible on the bushes right through winter and into the following spring. The fruits are usually ready to pick in late summer and fall, when they can be strung to dry – strings of them make pretty Christmas decorations. The remains of the flowering calyx usually persist as a cluster of brownish, pointed sepals at the end of the hip.

Wild roses grow readily in the garden and can be an attractive screen or back- 59 drop in a wild corner. The bushes provide excellent shelter for songbirds, and the blossoms can be breathtakingly beautiful. Rose seeds germinate slowly and may require winter stratification; cuttings or rooted shoots dug in fall or early spring are probably a more satisfactory means of propagating wild roses.

Rubus spp

(blackberry, dewberry, brambleberry)

There are about a dozen species of blackberry growing in open woods and thickets in parts of Canada and the northern United States. All have woody stems armed with sharp thorns or prickles. Those with more slender trailing stems are sometimes called dewberries, but any of them may be called brambleberries. The leaves are compound, usually with three to five pointed, toothed leaflets, often veined with spines beneath. The white to pinkish flowers are borne in clusters. The juicy, usually delicious berries, which ripen to dark red or black by late summer or fall, are actually composite fruits con-

sisting of numerous small, one-seeded drupelets attached to a whitish receptacle which, unlike that of a raspberry, detaches with the fruit when picked.

Wild blackberries make delicious pies, jams, jellies and wines and combine well with other fruits such as apples. Indigenous people made a fruity tea from the leaves, especially after they turned reddish in fall. The leaves of western trailing blackberry (*R. ursinus*), which make a particularly good tea, were used by the Straits Salish of Vancouver Island to sweeten herbal medicines.

Most blackberries can be propagated easily from cuttings or layered shoot tips (see Chapter Three).

Rubus chamaemorus

(cloudberry)

Sour and disagreeable to some people, cloudberries are relished by others. In Alaska, they are called salmonberries, a name that farther south refers to *R. spectabilis*, a much larger, prickly stemmed relative. In Newfoundland and elsewhere, cloudberries are marketed commercially and sometimes called baked apple or bake-apple, perhaps derived from the French *Baie qu'appelle*? (What is this berry called?). In Scandinavia, where large quantities are harvested, experiments are under way to initiate large-scale cultivation. The berries can be made into jam and other preserves or simply served fresh with cream and a little sugar or honey. A dish known as Inuit ice cream or Eskimo ice cream is made by beating cloudberries with a mixture of seal oil and chewed caribou tallow.

Cloudberry grows about 6 inches tall in moist, peaty soil throughout much of northern Canada and southward in acid bog sites, and its slender, unbranched stems bear only one to three round-lobed leaves. The flowers are white and solitary, with male and female produced on separate plants. The compound fruits have

taining more robust canes, larger fruits and higher yields, have never quite been able to capture the fragrance and flavour of the wild berry.

The wild raspberry grows in open woods and rocky areas and along streams from Newfoundland to British Columbia and south to North Carolina and northern Mexico. In Canada, it is absent only on the

relatively few large drupelets. Immature cloudberries are firm and reddish, becoming amber, then pale yellow, soft and very juicy when fully ripe.

Several additional *Rubus* species grow in various parts of Canada and the neighbouring states. These include Arctic raspberry or nagoonberry (*R. arcticus*), hairy raspberry (*R. pubescens*) and trailing raspberry (*R. pedatus*), all of which have edible red berries and lack spines or thorns. Cloudberry's soil requirements make it a good partner for blueberries and bog cranberries in a peat-bog garden with plenty of sphagnum moss.

Rubus idaeus

(wild raspberry)

The wild raspberry is very similar to its domesticated counterpart, with clustered, prickly canes, three-part deciduous leaves, white flowers borne singly or in clusters and delectable, juicy, aggregate fruits that fall into one's hand when fully ripe. In fact, cultivated raspberries originated from different strains of this wild type and are included within this species. In my opinion, the plant breeders who developed the cultivated forms, while successful in ob-

northern tundra and alpine slopes and in the Pacific rainforest. Many different forms are recognized, including one with pink flowers and one with amber-white berries. There are also several shrubby raspberry relatives, almost all with delicious fruits, such as the black raspberry, or blackcap (*R. occidentalis* and *R. leucodermis*), with white flowers and dark, purplish, juicy berries.

All of these species are wonderful in jams, jellies, fruit drinks or pies and other baked desserts. They can easily be frozen for winter and can also be dried individually or as fruit leather. The indigenous people of British Columbia once picked, peeled and ate the tender young sprouts 61 of salmonberry (*R. spectabilis*) and thimbleberry (*R. parviflorus*) as spring greens.

Wild raspberry and its relatives can be grown as landscaping shrubs in well-drained soil in open, sunny places. They can be propagated from cuttings or layered shoots as well as from seed. Salmonberry generally prefers a moister, shadier habitat than the others.

Sambucus spp

(elderberry)

Blue elderberry (*S. caerulea* or *S. glauca*) is a large, bushy, deciduous shrub well known for its small, blue, waxy-coated berries that follow masses of creamy flowers borne in rounded or flat-topped clusters. The leaves, in opposite pairs, are pinnately compound, each bearing five to nine pointed, oval leaflets with toothed margins. The berries, seedy and tart, grow in clusters that are often so dense they weigh down the branches; one can pick a bucketful in a very short time.

Blue elderberry grows in valleys and on open slopes, especially along watercourses from southern British Columbia to western Montana and southward to California and New Mexico. Red elderberry (*S. racemosa*) is similar but has pyramidal flower clusters and bright red (or occasionally whitish, yellow or shiny purple-black) berries. It is common and widespread across North America, but its fruits are considered inferior to those of blue elder. The fruits of red elder should always be cooked, since they are reputed to cause nausea if eaten raw. Common elderberry (*S. canadensis*), which grows in eastern and central North America, is also similar but has black-fruited berries.

Native people sometimes used elderberry juice for marinating fish or other foods. Homemade elderberry wine has a fine reputation, and elderberries make excellent jellies and can be cooked with apples and other fruits in pies and cobblers. The mature flower clusters are also edible and can be fried or cooked in muffins and pancakes, but the leaves, bark, stems and roots of elderberries are poisonous, as they contain a cyanide-producing compound.

Elderberries can be grown from seeds planted in the fall or from cuttings. Garden types can be purchased from nurseries. All are excellent shrubs for attracting birds and other wildlife.

Shepherdia spp

(buffalo-berry, soapberry)

Silver buffalo-berry (*S. argentea*), a deciduous shrub or small tree closely related to the exotic ornamental Russian olive tree (*Elaeagnus angustifolia*), can be recognized in the wild by the dense, greyish scales, or scurf, covering its leaves and twigs, giving it a silver lustre. The leaves are smooth-edged and the older branches usually tipped with spines. The small, inconspicuous flowers are borne in clusters in the axils of the leaves. The plant grows in open woods and thickets and on rocky slopes and shores throughout the southern Prairie Provinces and Rocky Mountains and south to California and Iowa.

The berries are small, oval, scarlet and so tart that they are almost inedible until sweetened by the first frost. The name buffalo-berry is apparently derived from the former practice of some Plains Indians of flavouring buffalo meat with a sauce made from the berries. More recently, the fruits have been used for jams and jellies.

A close relative, *S. canadensis*, known as soapberry, soopollalie (the Chinook name) or russet buffalo-berry, grows in open woods, on shores and in thickets across Canada and the northern United States. It is also a deciduous shrub but thornless, with twigs and leaves covered with coppery or rust-coloured scurf. Its berries are bright orange to red and extremely bitter. Nevertheless, they were and still are esteemed by indigenous people in British Columbia and neighbouring areas as the source of a favourite whipped confection known as Indian ice cream. The berries are whipped with water (and nowadays, a little sugar) into a pinkish froth that is eaten as a special dessert at feasts and family gatherings. (To make this dessert, whip 2 to 4 tablespoons of soapberries with about ½ cup of water and a little sugar. Note that if any oil or grease comes in contact with the berries, they will not whip.)

Both silver buffalo-berry and soapberry are interesting and attractive garden ornamentals in open, well-drained sites, especially in limy soils. They are best propagated from cuttings; the seeds must be scarified to soften the hard seed coats and stratified to break dormancy. Because male and female flowers grow on separate plants, the two types of bush must be interplanted for berry production. Occasional golden-fruited forms appear; 'Xanthocarpa,' for example, a yellow-fruited cultivar of silver buffalo-berry, is sold by some nurseries.

Vaccinium spp

(blueberry, huckleberry)

About 20 species of shrubs in the genus *Vaccinium* are native to Canada and the northern United States. Virtually all are prized in one region or another. In fact, they are probably the most favoured of all the wild berries. Each has its own characteristics and may be called variously wild blueberry, huckleberry, whortleberry or bilberry. They vary in size from diminutive dwarf blueberries (*V. caespitosum*) and bog blueberries (*V. uliginosum*) to the highbush blueberry (*V. corymbosum*), which may be taller than 6 feet.

The leaves are generally deciduous, except for those of evergreen huckleberry (*V. ovatum*). Most flowers are small, urnshaped and whitish or pinkish. The berries may be single or clustered, and all are globular, juicy and flavourful. Most of the types called blueberries have blue fruit, usually with a waxy bloom on the skin. There are also red-fruited species, such as the red huckleberry of the Pacific Coast (*V. parvifolium*), and black-fruited kinds, such as black mountain huckleberry (*V. membranaceum*), an exceptionally deli- 63

seasonal gathering round. Today, its shiny, dark green foliage is used in commercial floral arrangements.

The vacciniums may be propagated from seeds, cuttings or rooted shoots, and like their domesticated counterparts, they prefer damp, slightly acidic soil.

Vaccinium vitis-idaea

(lingonberry, lowbush cranberry)

cious western type, considered by some native people to be the best of all the berries. In addition to the huckleberries in the genus *Vaccinium*, there are two in a separate but related genus, *Gaylussacia*.

Blueberries and huckleberries grow in a wide variety of habitats, from coastal thickets and woods to peat bogs and alpine slopes. Most prefer an open, sunny location with plenty of soil moisture. Indigenous people sometimes practised controlled burning of mountainsides in order to maintain an optimum habitat for blueberries and other desirable food plants. In fact, recent studies by the United States Department of Agriculture indicate that wild blueberries and huckleberries are dwindling in the Pacific Northwest because forestry practices discourage burning as a method of habitat enhancement.

Blueberries and their relatives generally ripen in summer and produce abundantly. Some people use a comblike implement for harvesting them. Evergreen huckleberry, sometimes called winter huckleberry, ripens later than the others, producing small, clustered, black or powder-blue berries. For the northwest-coast people of Vancouver Island, it was the last type of berry to be picked on the

Lingonberry, a close relative of blueberries and bog cranberries, is a dwarf, evergreen, mat-forming shrub, seldom taller than a foot, that grows on peat bogs, rocky tundra and barrens, although it usually produces no fruit above the tree line. It is known by various other names, including mountain cranberry, European cranberry, rock cranberry, cowberry, foxberry and partridgeberry (this last name is also applied to an unrelated creeping plant of eastern North America, *Mitchella repens*, which produces edible red berries in pairs). Lingonberry leaves are small, shiny, rounded and leathery, resembling those of kinnikinnick. The pinkish white, urn-shaped flowers are borne in small clusters at the tips of the tufted branches. The berries are bright red and usually

64

quite small, but they are very attractive. The plant is widespread in northern Europe, Asia and North America, where it grows from the Arctic to central British Columbia, to Lake Superior in Ontario, east through Quebec and the Atlantic Provinces and into New England.

Lingonberries ripen in late summer and fall and may persist on the plants throughout the winter. Somewhat tart and hard at first, they become softer and much sweeter and tastier after exposure to frost or storage in water. The berries make excellent jam, and like bog cranberries and their commercial counterparts, they can be cooked into a delicious sauce, said to be superior in flavour to commercial products. They were and still are a very important fruit for indigenous people in northern North America and Europe.

In the garden, lingonberry requires moist, acidic soil. It can be propagated from cuttings, layered shoots or seed. Fortunately, in many areas, it is sold as a landscaping ornamental. It provides an attractive, interesting ground cover, with the added advantage that it produces excellent fruit. Wild birds also relish the berries.

Viburnum edule

(highbush cranberry, squashberry)

Highbush cranberry, sometimes confused with the low-growing cranberry of the genus *Vaccinium*, is an erect deciduous shrub in the honeysuckle family. It grows as tall as 8 feet, with smooth, reddish bark and opposite leaves that usually have three lobes. The small white flowers grow in rounded clusters, and the orange to bright red fruits are globular and shiny, each containing a single flattened seed. When unripe, the fruits are hard and very sour and may give off an unpleasant musty odour, but after they have been touched by frost, they become softer and more palatable. Indigenous people sometimes leave highbush cranberries on the bush well into winter. A common method

of storing the berries was to place them in containers and cover them with water or a mixture of water and vegetable oil that would gradually soften and sweeten them.

This shrub is found in moist woods and thickets across Canada and the northern United States. Highbush cranberry grows well in moist, acidic soil in partial shade and can be propagated from cuttings or from layered or rooted shoots. Several other species of viburnum, all with edible fruits, grow throughout various parts of the continent. Snowball bush, or Guelderrose, a popular flowering shrub, is related to highbush cranberry, but it is usually sterile and therefore lacks fruits.

Poisonous and Unpalatable Berries

Baneberry, red- and white-berried (*Actaea rubra* and *A. pachypoda*) – Berries and entire plants of both forms are very poisonous.

Bittersweet (*Celastrus scandens*) – Berries are disagreeable and reputed to be poisonous.

Buckthorns (*Rhamnus* spp) – Berries have purgative properties.

Cascara (*Rhamnus purshiana*) – Berries have purgative properties.

65

The entire baneberry plant, including its white or red fruit, is very poisonous.

Cherries, wild (*Prunus* spp) – Flesh is edible, but pits can be poisonous if swallowed in quantity. Foliage and twigs are also poisonous.

Cohosh, blue (*Caulophyllum thalictroides*) – Berries are bitter and poisonous.

Comandra or bastard toadflax (*Comandra umbellata*) – Berries are edible but bland, dry and seedy.

Daphne (*Daphne mezereum*), daphne-laurel (*D. laureola*) – Berries and foliage are very poisonous.

Devil's club (*Oplopanax horridus*) – Berries are reputed to be poisonous, and they have spines.

Dogwood (*Cornus* spp) – Berries are unpalatable.

Fairybells (*Disporum* spp) – Berries are edible but seedy and insipid.

Geocaulon or timberry (*Geocaulon lividum*) – Berries are questionable and not generally eaten.

Hawthorns (*Crataegus* spp) – Berries of most species are edible but may be seedy and insipid. May cause constipation.

Holly (*Ilex* spp) – Berries are inedible and mildly poisonous.

Honeysuckle (*Lonicera* spp) – Berries of most species are considered inedible.

Jack-in-the-pulpit (*Arisaema* spp) – Berries and entire plants are acrid and inedible.

Juniper (*Juniperus* spp) – Strongly flavoured berries are used in gin and as a seasoning for game and stews, but they should not be eaten in quantity and not at all by pregnant women, as they are known to cause uterine contractions.

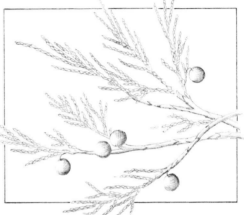

Juniper berries flavour gin, but all species should be used with caution.

Lily-of-the-valley, wild (*Maianthemum* spp) – Berries are edible but not palatable. (The berries and plants of domesticated lily-of-the-valley, *Convallaria majalis*, are very poisonous.)

Mayapple (*Podophyllum peltatum*) – Fully ripe fruits are edible and of good flavour, but unripe fruits, leaves and roots are very poisonous.

Mistletoe (*Phoradendron* spp, *Loranthus* spp and related species) – The whitish, sticky berries and the entire plant are poisonous.

Moonseed (*Menispermum canadense*) – Berries are very poisonous and potentially fatal.

Mountain ash (*Sorbus* spp) – Berries are edible but very bitter. Seeds should not be consumed raw in quantity.

Nightshades (*Solanum* spp) – Berries (except of some cultivated forms) are bit-

The ornamental mountain ash produces berries that are edible but very bitter.

Twinberry or twinflower honeysuckle (*Lonicera* spp) – Berries are generally considered inedible but are apparently not highly poisonous.

Twisted-stalk, cucumberroot or watermelon berry (*Streptopus* spp) – Red, translucent berries are edible but seedy.

Mayapples turn from poisonous when unripe to edible and delicious when ripe.

ter and poisonous, especially when unripe. Potentially fatal.

Poison ivy, poison sumac (*Toxicodendron* spp, *Rhus* spp) – Berries should be left alone. Plants can cause severe allergic reactions. The small reddish berries of the related sumacs (*R. glabra* and *R. typhina*) are edible, though tart, and can be used to make a lemonadelike beverage.

Pokeweed or pokeberry (*Phytolacca americana*) – Raw berries and seeds are very poisonous – potentially fatal. Plants contain blood-cell-destroying lectins and should not be handled.

Privet (*Ligustrum vulgare*) – Berries are very poisonous – potentially fatal.

Queenscup (*Clintonia uniflora*) – Berries are generally considered inedible.

Silverberry (*Elaeagnus commutata*) – Edible but dry, seedy and insipid fruits.

Solomon's seal, false and star-flowered (*Smilacina* spp) – Berries are edible but small and seedy.

Spindlebush or wahoo (*Euonymus* spp) – Bright orange fruits and entire plants are poisonous.

Virginia creeper (*Parthenocissus quinquefolia*) – Purple, grapelike berries and entire plants are poisonous.

Yew or ground-hemlock (*Taxus* spp) – Wood, bark, needles and seeds within the "berries" are highly poisonous. The flesh surrounding the seed is edible but best left alone.

Nancy J. Turner of Victoria, British Columbia, is a recognized Canadian authority on ethnobotany, the study of traditional uses of plants and their importance to indigenous peoples. She co-authored a series of books on wild plants published by the Canadian National Museum of Natural Sciences.

67

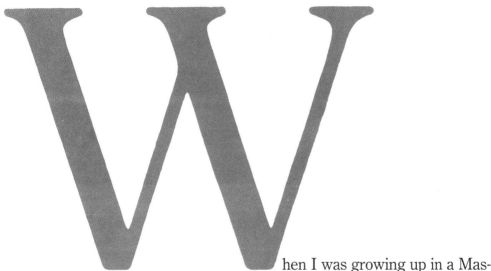

When I was growing up in a Massachusetts suburb in the 1930s and 1940s, much of the fruit I ate was canned in a heavy syrup. It came straight from the brand-new supermarket around the corner. No wonder one of my most vivid memories of those years is of an entirely different experience: picking and eating wild raspberries along a dusty back road of rural Maine, where I often spent summers as a camper. I see myself clearly, even from a distance of more than 40 years. I am 9 or 10, my hair in pigtails, wearing a regulation green cap, a white cotton short-sleeved shirt and regulation camp shorts—green with white stripes on the sides—bobby socks, brown and white saddle shoes and a sweater tied around my hips. One sweaty hand clutches a brown paper bag lunch (two peanut butter and jelly sandwiches and a large red apple), while the other reaches out to pluck the irresistibly plump, sun-drenched fruits growing in profusion along the roadsides. Wild raspberries! What a revelation to a child of the suburbs who thought all fruit originated in the local Stop & Shop. I could not have known that this experience would lay the foundation of my lifelong interest in small fruits, especially wild and cultivated berries. I could not have known, in those long-ago summers by Little Sebago Lake, that one day I would live in the backlands of Cape 69

Both cultivated berries and edible wild species picked in quantity can be cooked into a delicious variety of jams, jellies, sauces, fruit butters, leathers, juices and wines.

Breton Island, in northeastern Nova Scotia, on an old-fashioned, self-sufficient farm, where I would learn to preserve the fruits of my family's labours in jams, jellies, sauces, fruit butters, leathers, juices and wines. Fruit growing, harvesting and preserving have become, over time, very important features of my life.

In the early years of my marriage, long before my husband Jigs and I owned our own farm in Cape Breton, we picked and preserved the wild berries we discovered wherever we lived. We moved around a lot—15 times in 16 years—and with a growing family and a tiny income, we had to learn how to enrich our diet by harvesting wild fruits and other edibles. These early experiences were central to our evolving self-reliance.

Each successive move took us deeper into the countryside, into ever more remote areas. We began to acquire rudimentary farm accoutrements—a small

vegetable garden, a few chickens in the backyard, a cow in the garage. The kitchen provided an archetypal scene of rural living: great pots of berries or jelly or jam simmering on a wood-burning cookstove, stirred occasionally with a long-handled, homemade wooden spoon.

My humble role in these undertakings was as sous-chef, assistant to the head cook. My job was to find the right-size jars for the fruit at hand, sterilize them—I became adept at fishing hot jars out of boiling water with a long-handled fork—and seal them. I topped jams and jellies with melted paraffin, making a perfect seal by tilting the jar so that the hot wax curled upward, sealing the entire inside edge of the jar. Just two thin layers, and the preserves could be kept in mint condition from one season to the next if stored in a cool, dark place—usually our cellar, sometimes a pantry. How wonderful it was to open the bright jars in the dead of winter and taste once more the intense flavours of homegrown berries.

When we made jelly, wine or juice, Jigs boiled the berries, then drained them in a cheesecloth jelly bag suspended from an old broom handle balanced between two chairs, just as he'd seen his mother do. The process is called extracting. Very nice but expensive steam juicers that eliminate the need for Rube Goldberg broom-handle extractors are now on the market, but we stick to the old way that has served us well. Though we have often varied the procedure – I now hang the bag from a bolt on the end of our long wooden trestle table – the principle of the hanging jelly bag is unchanged. We feed the resulting dry berry mash to pigs if we have them; otherwise, it goes on the compost heap. With strongly flavoured fruits such as black currants, we can make three or four extractions from the same berries. Jigs makes a lot of black currant wine this way from a small amount of fruit.

One of the jellies we make is spicy and can be used as a condiment for venison or meat dishes. It can be made with grapes, but I have altered it to accommodate elderberries, a more familiar fruit. (This recipe and those on the following pages have been adapted from my book *The Old-Fashioned Fruit Garden* with the permission of Nimbus Publishing Limited of Canada.)

Spiced Elderberry Jelly

6 lbs. apples, cut up
4 qts. ripe elderberries with stems
1 qt. cider vinegar
1 qt. water
1 Tbsp. ground cinnamon (or ¼ cup cinnamon stick)
1 Tbsp. ground cloves (or ¼ cup whole cloves)
4 cups sugar

Combine apples, elderberries, vinegar and water in a large preserving pot. Cover and bring to a boil, simmering until all the fruit is soft. Drain the hot mixture through a jelly bag for several hours or overnight. Squeeze the bag to extract all the juice. Measure 1 quart of juice into a pot, bring it to a boil, and stir in the spices (if whole, put them in a little cheesecloth bag) with the sugar. Boil, uncovered, for about 15 minutes or until a small amount sheets off a metal spoon. Remove the spice bag, pour jelly into hot jars and seal.

Raspberry Juice

We make a delicious raspberry juice by mashing 6 quarts of ripe or overripe berries in a large pot, adding a little water and simmering, stirring frequently, until the juice runs freely. Let it drip in the jelly bag for several hours or overnight. To each quart of juice, stir in ½ to 1 cup sugar. Bring to a boil, reduce heat, and simmer for 5 minutes. Pour into hot, sterilized jars, and seal at once with snap lids and screw bands. Substituting currants for some of the raspberries produces a distinctive, refreshing flavour.

Jigs' first teaching job took us to the eastern shore of Maryland, where we lived in a large, creaky house at the edge of the Chester River. We furnished it, as I recall, for $90 – including carpets, beds, bureaus, sideboards, a large dining-room table and silverware – from the Salvation Army. Our worldly goods consisted mainly of books (including a few treasured cookbooks) and an increasing accumulation of hand tools and old-fashioned kitchen equipment.

Here, my husband learned to hunt wild ducks, geese and game of all kinds, and I learned to prepare them for the table. We expanded our vegetable garden, but our fruit still came from the wild.

The most abundant wild berries were blackberries, and what 'huge, magnificent fruits they were. A handful was enough for a satisfying bowl of berries and cream, and though seedy, they could 71 be made into jam without straining because the fruit was so meaty that the seeds were hardly noticeable.

I discovered that great quantities of overripe berries could easily be turned into syrup that had a variety of uses. It could be poured over pancakes (preferably light ones made with buttermilk), French toast or ice cream. Diluted, the syrup was transformed into juice or used

Many jellies require additional pectin, easily made from any tart, juicy apples.

as the base for jellied fruit desserts. I was beginning to learn how to take advantage of what each fruit had to offer and to preserve and use it in the simplest and most efficient ways. With four small children and a very limited income, I had neither the time nor the money to use fruit frivolously. Far from feeling that this was a disadvantage, I enjoyed learning the art of "making do," of making something out of nothing (or almost nothing) and still producing fruit products that family, friends and guests enjoyed.

Blackberry Syrup

To make blackberry syrup, mash fully ripe fruit (as much as conveniently fits into a preserving pot), add water to barely cover the fruit, and bring it to a boil. Sim-

mer the berries for about 10 minutes, stirring often, and when the juice is running freely, strain it through a jelly bag for several hours or overnight. (If you have a steam juicer, follow the manufacturer's directions.) Add ½ cup of sugar for each cup of juice, and boil, uncovered, for 10 or 15 minutes, until the mixture thickens. Then pour it into hot jars, and seal at once with snap lids and screw bands, tightening all the way. It does not need further processing.

Years later, our youngest child, Curdie, gave us a wonderful recipe for blackberry pie, which is good anytime during summer when the berries are plentiful. Ingenious and simple, it is my kind of dish.

Curdie's Blackberry Pie

2 cups sugar
1 cup water
1 qt. blackberries
Baked 9-inch pie shell
Whipped cream

In a pot, boil together the sugar and water, uncovered, to form a syrup, which takes about 5 minutes. Stir until the sugar is dissolved. Pour the hot syrup over the blackberries, and place the mixture in the pie shell. Top with whipped cream, and refrigerate until ready to serve.

A word about whipped cream. It helps to have at least one Jersey cow around, as we have had for almost 30 years – Jerseys produce the best cream. It is easiest to skim off the milk when it is chilled and fresh, preferably no more than 12 hours old. The bowl and beater should be well chilled. I use an old-fashioned hand mixer or beater with a tight-fitting lid that holds the beater firmly (no splashing cream). Most people do not have their own cow, but real cream is still a must. No extra sugar and no vanilla need be added; the real thing needs no gussying up. If you can manage to keep a Jersey, you can make a number of high-quality cream products

"During the season, we look forward to the ripening of each type of berry. How delight- *ful it is to savour each fruit and preserve its special flavour for the dark days of winter."*

(ice cream, Devonshire cream, cream cheese, cottage cheese, yogurt) that are elegant, no-fuss accompaniments to just about any fruit.

In 1962, we made a decision to follow a path we continue to travel today, one that has been called Simple Living, though as we discovered from experience, it is certainly not simple. We moved to a 50-acre farm on a hillside in northern Vermont in the fall of that year – no plumbing, no electricity, rent $10 a month. Our assets included four small children, $300 in savings, a cow named Aster, a gorgeous view of the valley below and the enthusiasm and energy of youth.

The first winter, all our vegetables froze in the cellar. So did the gravity-fed water. And we ran out of wood. The owner of the farm, then living in New Zealand, had extolled the virtues of Simple Living but failed to explain (in a booklet he had written) that he and his family spent the hardest part of the winter in the south, that the family automobile was not mentioned in his budgetary accounts and that 73 he did not really do his family's laundry in the nearby bubbling brook but in a neighbour's washing machine.

Experience, they say, is a dear school, and fools will learn in no other. This two-year sojourn put steel in our souls, enabling us to survive the difficult years ahead on a very small income as we continued to seek a way of life that would allow us to combine our farming skills with teaching, writing, thinking and the values of Simple Living.

Some of the wild fruits we especially enjoyed harvesting and preserving in the 1960s in Vermont were chokecherries, mountain cranberries and barberries. Who but my inventive husband could have discovered the uses of wild common barberries? Recently, when I checked the species—there are several—and asked Jigs which one he used to pick for preserves, he said, "I have no idea. I never knew the Latin name. I was only interested in feeding my family, and I picked the berries of the wild kind." Wild common barberries (*Berberis vulgaris*) were brought to the New World by the early settlers and now grow as escapees in southern Ontario and Quebec and all over New England, naturalized along back roads and banks wherever conditions are favourable. The brilliant orange-red, oval fruits, which hang in long clusters, ripen in the fall. This species is also one of the most ornamental, with long, arching branches and colourful fall foliage. Common barberries, unfortunately, are hosts of the black rust fungus that affects wheat. The berry is not usually sold as an ornamental shrub, so the fruits are only available in the wild.

Barberry Candy

The barberries were very simple to prepare. The basic recipe was gleaned from an old cookbook, the original source of many of our fruit recipes. Measure berries and molasses into a pot in equal amounts

Highbush cranberries can be used for a sauce suitable for Thanksgiving dinner.

by volume, cover, and bring to a boil. Skim out the berries, and reboil the syrup, reducing it by half (be careful that it doesn't burn). Add the berries, bring the mixture to a rolling boil, and ladle it into hot, scalded jars. Seal them at once. The result is a somewhat crunchy, sticky confection that tastes like licorice. I confess I never liked it, but our children did. To serve, just open the jar and pass it around with spoons—and plenty of napkins.

Highbush cranberries (*Viburnum trilobum* or *V. opulus americanum*) are common in the northern United States and southern Canada. We picked ours along the dirt roads of northern Vermont. The bush is ornamental in every season. First there are flat clusters of white flowers in late spring, then shiny red fruits and brilliant foliage in the fall. The fruits are acid and loaded with pectin, like the better-known bog cranberry (*Oxycoccus* spp), so they make an acceptable substitute for

Thanksgiving cranberry sauce. We made the fruit into jelly to spread on bread.

Highbush Cranberry Sauce

Just simmer a quart of berries in 2 cups of water until the skins crack. Put the fruit through a hand food mill, a blender or a food processor to mash it, then add 2 cups of sugar. Bring the mixture to a boil, and cook it for about five minutes, uncovered. Pour the sauce into hot, scalded jars, and then seal with either wax or snap lids and rings.

Chokecherries also grew in profusion along backcountry roads in Vermont. These, we learned, could not be made into jelly unless some form of pectin was added to them. At first, my husband used commercial pectin, but it requires a lot of sugar, which masks the fresh, tart flavour of chokecherries (indeed, of any fruit). Eventually, he learned to make his own pectin with wild apples.

Homemade Pectin

Chop 2 pounds of tart, juicy apples, including peels and cores. Place in a large pot with a quart of water, cover, and simmer until the apples are soft, stirring occasionally. Strain through a jelly bag overnight. The next day, boil hard, uncovered, for 20 minutes. Stir in 2 tablespoons of lemon juice, bring to a boil again, and pour into hot, sterilized jars. Close with snap lids and screw bands, and process for 5 minutes. This recipe makes about 2 pints of a concentrated juice that will help set any fruit. During the 1960s, Jigs, whom we dubbed "The Jelly King of the Northeast Kingdom," began to sell chokecherry and other jellies on a modest scale.

By 1970, we had made the move with a Noah's ark of animals — a team of workhorses, two Jersey heifers, a small flock of chickens — to our first farm, our very own piece of land on Cape Breton Island. Here, we began to farm in earnest, free at

Chokecherries cannot be made into jelly unless some form of pectin is added.

last to cultivate the fruits that thrive in a northern climate: strawberries, raspberries, gooseberries, black and red currants, blueberries, elderberries. We gave up our old truck (or rather, *it* gave up), because we knew that the poor northern land could not grow enough to enable us to pay for it. With no vehicle and the nearest city 58 miles away, we had to learn to raise just about everything we needed on the farm.

Our hard work in improving the soil paid off in abundant crops of berries, some of which we sold — 75 cases of jam and jelly in a summer — and the remainder of which we preserved in canning jars in a light syrup (we had no freezer until quite recently) or turned into other fruit products. When the fruit was in season, of course, we and our guests enjoyed pure and simple fresh berries and cream three times a day.

With this move came a radical reorganization of labour. My husband was no 75

In July, when the elderberries bloom in Nova Scotia, the author adds the sweet, cream- coloured florets to pancake batter, replacing a half-cup of flour in the recipe.

longer free to indulge in his favourite pastime of preserving, since there was other work that came first: ploughing, planting and harvesting with the team of horses, woodcutting and haying. I took over all his former tasks, including cheese and butter making as well as fruit preserving, and as the children left for the wider world, I also learned to make a square load of hay on the big old hay wagon.

In the kitchen, I began to rationalize my efforts. Gooseberries don't really need to be topped and tailed. What a revelation! Especially when they have been harvested by the gallon.

My most earthshaking discovery was that Jigs' approach—large pots of simmering fruit—does not produce the best results. The world's best jams and jellies are quickly cooked in small batches, only 1 quart of fruit or juice at a time. I learned that a wide-mouthed, 2-gallon, stainless steel pot, while not as appealing as a big old enamel preserving pot, is worth its weight in gold in jam and jelly season, for it encourages the fast evaporation and evenly distributed heat that best preserve each fruit's unique flavour. I discovered, in short, the never-fail guidelines that have enabled me to turn our fruit harvests into excellent products.

I did not invent these first principles of jam and jelly making. They have been known for ages but have been obscured by recipes for freezer jam and boxes of commercial pectin, which take both skill and flavour out of home preserving.

The World's Best Jam

If you want to make the World's Best Jam, just follow these steps: Pick a mixture of just-ripe and underripe berries, then measure a quart of berries and lightly mash them. Heat them to simmering, and stir in the required amount of sugar. Bring the mixture to a boil, and simmer, stirring occasionally until it jells.

This should take no more than 10 or 15 minutes. If your berries do not make jam in this time, forget it. Turn them into a sauce, and don't be ashamed. It will be delicious on pancakes or ice cream, far more delicious than it would be if you cooked the mixture into treacle. The secret of success is to make use of each fruit's natural pectin and to use only enough sugar to make it set. A basic rule of thumb is to use equal amounts by weight of fruit and sugar. And remember that undercooked jam or jelly will develop moulds, while overcooked jam or jelly obscures the fruits' distinctive flavours.

Of course I made mistakes, but virtually every one of them was turned to advantage. My runny jams and jellies, with a magic wand, became delicious fruit sauces we called "sass," a variation of the old New England (and English) term for cooked vegetables, "garden sass." Sass, served with feather-light buttermilk pancakes, became the focus of Sunday-morning guest breakfasts.

I deliberately began to make sass from a variety of fruits, working out the sugar/juice proportions to produce tasty, not overly sweet, strawberry, raspberry and elderberry sauces. I also began to make crocks of brandied fruit when the strawberries and other fruits were especially plentiful: 1 quart of brandy and 2 cups of sugar for each quart of fruit – what could be simpler?

During elderberry blossom season – July, here – I add the sweet, cream-coloured florets to pancake batter, substituting them for half a cup of flour in the recipe. (Elderberry flowers can be used as a partial flour substitute in muffin batter too.) Then, what a breakfast! The different sasses are lined up on the table with brandied fruit, a dish of fresh Jersey butter, a bowl of naturally soured cream nearby and even a pitcher of maple syrup for form's sake.

Except for strawberries and raspberries, which require a lot of hand weeding, our berries are easy to care for (especially the chokecherries the birds have so thoughtfully planted for us along our half-mile lane). Everything is thoroughly weeded in the spring, then mulched with three layers of material: compost on the bottom, then a layer of paper, old grain bags, worn-out carpet, whatever, topped with a layer of eelgrass or sawdust.

The poorest northern soil, we have discovered, will support berries of many kinds. Currants and gooseberries actually like damp, heavy soil. Recently, we have begun to plant strawberries and raspberries in raised beds lined with woven plastic bags (the type used for feed grain) topped with 6 to 8 inches of soil that we make ourselves from composted manure, sawdust and wood ashes. These more concentrated plantings do well in our poor growing conditions, virtually eliminating weeding.

During the growing year, we look forward to the ripening of each type of berry, and each one in turn becomes our favourite during its season in the sun. How fickle we are, yet how delightful it is to savour each fruit and preserve its special flavour for the darkest, coldest days of winter.

By learning to handle fruit with care and intelligence, one can raise, in the smallest backyard, most, if not all, of the fruit for an average family for a year. Such an undertaking gives some flesh to the notion of living off the land, though few northerners, and we are no exception, would seriously claim to be doing so. In any case, raising and processing one's own fruit is a very satisfying experience, one that I have pursued for many years.

Jo Ann Gardner, a contributor to *Harrowsmith* magazine, is the author of *The Old-Fashioned Fruit Garden* (Nimbus Publishing Limited, 1989). She grows and preserves a wide variety of berries in Cape Breton Island, Nova Scotia.

CLIMATIC ZONE MAPS – CANADA

Lower zone numbers refer to increasingly cold areas, but there are no specific minimum-temperature limits for each zone.

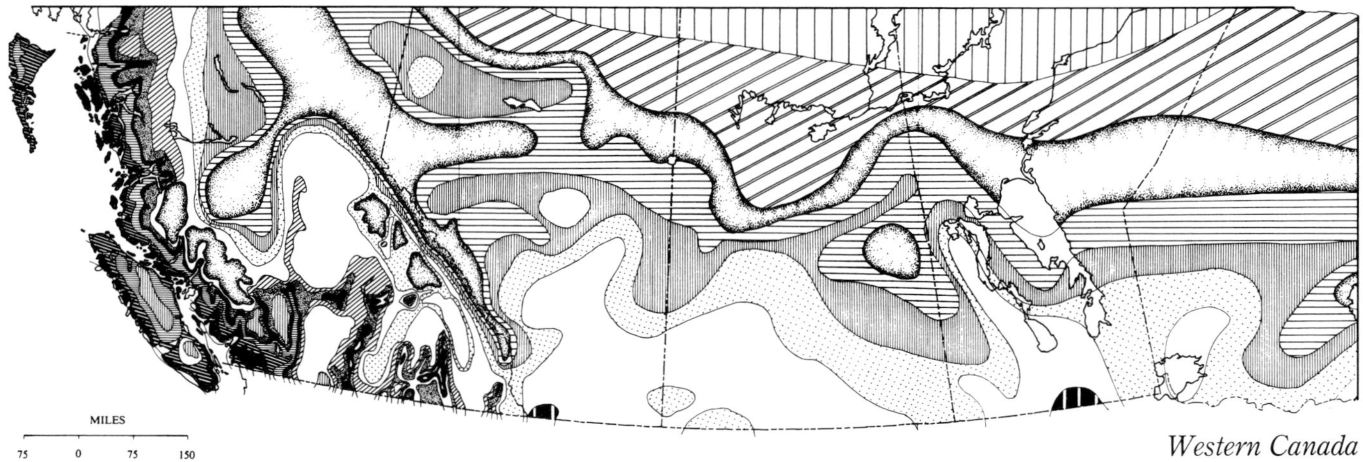

0a	
0b	
1a	
1b	
2a	
2b	
3a	
3b	
4a	
4b	
5a	
5b	
6a	
6b	
7	
8a	
8b	

Western Canada

MILES

75   0   75   150

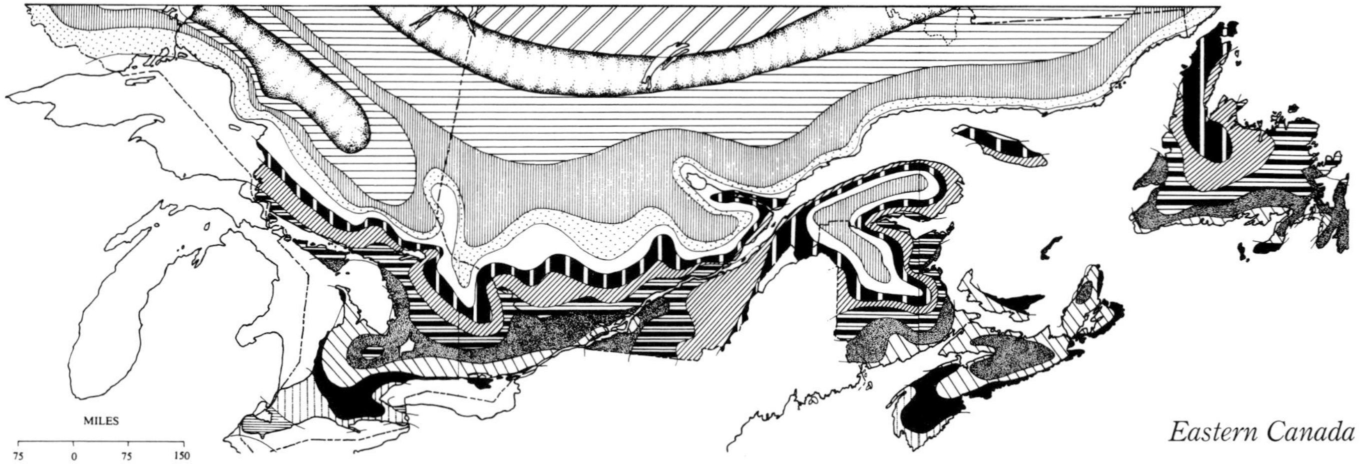

Eastern Canada

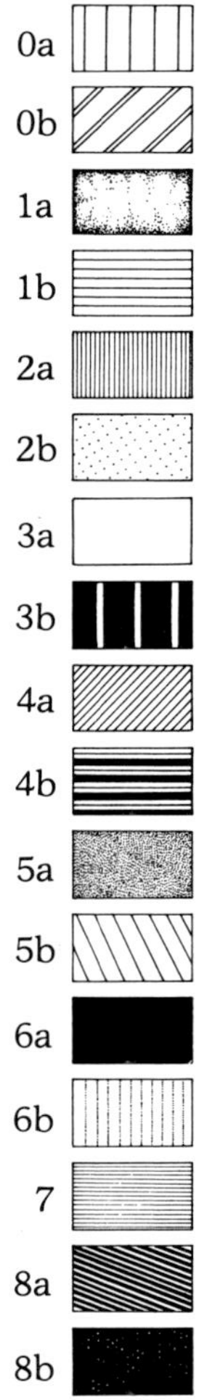

MILES

75   0   75   150

CLIMATIC ZONE MAP – UNITED STATES

Average minimum temperatures are listed for each zone in Fahrenheit degrees.

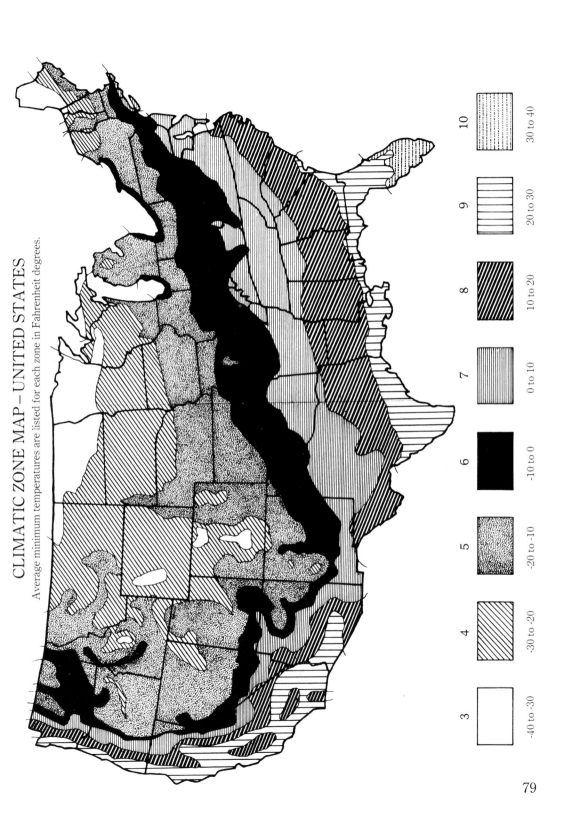

10	30 to 40
9	20 to 30
8	10 to 20
7	0 to 10
6	-10 to 0
5	-20 to -10
4	-30 to -20
3	-40 to -30

Achene
A dry, one-seeded **fruit** that does not split open along a well-defined line. Strawberries are composed of many joined achenes, and rose hips are filled with achenes.

Acidic
Having a **pH** below 7. Some berries, notably blueberries, require acidic soil. Soil can be made more acidic by adding sphagnum peat moss, sawdust or sulphur.

Alkaline
Having a **pH** above 7. Soil can be made more alkaline by adding ground limestone to it.

Axil
The angle between a leafstalk and the stem from which it grows.

Bacillus Thuringiensis
A bacterium, called Bt for short, that kills moth and butterfly larvae, caterpillars of the insect family Lepidoptera. As the bacterium is selective in its toxic effects, it is considered a safe insecticide by most organic gardeners.

Berry
A fleshy fruit, usually globular in shape, that encloses many small, scattered seeds. Currants, blueberries and gooseberries are thus true berries, as are cucumbers, grapes and tomatoes. Many fruits that are not technically berries, however, are popularly known as berries, as are a number of **drupes**, **pomes** and **composite fruits** described in this book.

Blight
A word that describes any plant disease, 81

although it is best used to describe diseases caused by fungi.

Bract
A modified leaf lying at the base of a flower or forming part of the flower. Ground-cherry fruits are enclosed by papery bracts.

Calyx
The outer series of modified petals in a flower head, composed of several **sepals**, usually green. It is the calyx that remains on the base of a strawberry or rose hip when it is picked.

Composite Fruit
A fruit made up of a cluster of smaller fruits joined together, as are raspberries and blackberries.

Compound
Compound leaves, as opposed to whole leaves, are made up of several small leaflets joined at the base or in predictable arrangements.

Cultivar
A plant variety that originated in cultivation rather than in the wild. 'Kent' is a strawberry cultivar.

Cutting
A piece of living plant – stem, leaf or root – taken from the mother plant and used to grow a new plant.

Deciduous
Describes a perennial plant that loses its leaves in fall. The opposite is evergreen.

Double
In describing flowers, refers to those with extra petals, usually as a result of breeding for ornamental qualities.

Drupe
A fruit that consists of a fleshy layer surrounding a hard stone or stones protecting the seeds within. Plums, cherries and elderberries are drupes, as are the individual fruits, usually called drupelets, that make up the **composite fruit** of a raspberry, cloudberry or blackberry.

Floricane
Raspberry or blackberry canes that produce flowers and fruit. In most **cultivars**, **primocanes** become floricanes in their second season.

Fruit
Botanically, any seed-bearing organ of a plant, but in common parlance, the term refers only to fleshy seed-bearing organs.

Genus
A group of similar living organisms within a larger family. The genus (plural genera) is divided into one or more **species**. When written, the genus name is italicized and begins with a capital letter. For instance, the plant *Rosa rugosa* belongs to the genus *Rosa*.

Herbaceous
Describes a plant that lacks a woody perennial stem.

Manure Tea
An organic fertilizer made by mixing about a quart of manure into a bucketful of water and letting the mixture sit for a couple of days before use.

Mulch
A soil covering meant to suppress weeds, hold in moisture and, depending on the cover, keep soil warm or cool. Mulches may be organic, consisting of leaves, straw, bark chips and such, or inorganic, made of plastic sheeting, old carpeting or special landscape blankets.

Palmate
In describing leaves, refers to those which are lobed like the fingers of an open hand. Currants have palmate leaves.

PH

Literally, potential hydrogen, a logarithmic measure of acidity or alkalinity in which the number 7 represents neutral. Lower numbers are increasingly **acidic**; higher numbers (to 14) are increasingly **alkaline**. Good garden soil is generally slightly acidic, with a pH of around 6.5.

Pinnate

In describing leaves, refers to those with a central stalk with leaflets extending from it on both sides in the form of a feather. Elderberry leaves are pinnate.

Pome

The fruit of saskatoon, apple, pear and similar plants. A pome is formed below rather than within the flower.

Primocane

The leaf-producing canes of raspberries or blackberries. Canes that produce flowers and fruit are called **floricanes**.

Receptacle

The central part of the flower, composed of the thickened end of the stalk, that holds all the floral organs in place. The receptacle stays within the fruit when a blackberry is picked but remains on the bush when a raspberry is picked.

Rotenone

A botanical pesticide made from certain tropical plants. This substance kills a wide range of pests and can be harmful to the user as well as to fish and amphibians. Its chief advantage is a short life span of toxicity in the garden, where it biodegrades into harmless substances.

Scarify

To scrape a seed with sandpaper or a knife so that it will germinate more quickly. This is done only to large seeds with hard coats, such as those of buffalo-berry.

Scurf

Minute scales on the surface of a plant that give it a silvery appearance.

Sepals

Small, modified leaves in a ring just outside the petals of some flowers. The sepals usually protect the flower bud. Together, the sepals form the **calyx**.

Single

In describing a flower, refers to one with the normal number of petals for the wild species. Those with additional petals are called **double** or semidouble.

Soil Test

A test that determines soil **pH** as well as the presence of specific nutrients. Most provincial departments of agriculture and state extension services can perform the test, or gardeners can do it themselves using soil-test kits available through some seed catalogues.

Species

The next smaller unit of classification of living organisms after **genus**. Members of the same species consistently breed true, and their names are written in lowercase italics. *Rosa rugosa* has the species name *rugosa*.

Stratify

To prepare seeds for germination by treating them to a period of exposure to cold and dampness. Only certain seeds, especially those of wild species whose seeds normally endure winter before sprouting, will benefit from stratification. Currant seeds, for instance, require stratification.

Sources: Mail-Order Directory

Mail-order nurseries often stock cultivars that are unavailable locally. While it is easiest and fastest to order plants from within one's own country, Canadians wishing to buy from U.S. nurseries can obtain a Permit to Import from The Permit Office, Plant Health Division, Agriculture Canada, Ottawa, Ontario K1A 0C6. One form is necessary for each company from which you order plants.

Imports to the United States must include an invoice showing the quantity and value of the plants, as well as a document from the Department of Agriculture certifying that the plants are disease-free.

Ahrens
RR 1
Huntingburg, Indiana 47542
Strawberries, blackberries, raspberries, currants, gooseberries. Catalogue free.

Bear Creek Nursery
Box 411
Northport, Washington 99157
Good selection of currants. Also cranberries, mulberries, gooseberries, elderberries, 'Jostaberries,' raspberries, saskatoons, blueberries, huckleberries. Catalogue free.

Blueberry Hill
RR 1
Maynooth, Ontario K0L 2S0
Blueberries. Brochure free.

Boughen Nurseries
Valley River Ltd.
Box 12
Valley River, Manitoba R0L 2B0
Hardy raspberries, currants, gooseberries, cranberries, saskatoons. Catalogue free.

Brittingham Plant Farms
Box 2538
Salisbury, Maryland 21801
Very good selection of strawberries; also blackberries, blueberries, raspberries. Catalogue free.

W. Atlee Burpee & Co.
Warminster, Pennsylvania 18974
Blueberries, blackberries, boysenberries, elderberries, gooseberries, grapes, raspberries, strawberries. Catalogue free, to U.S. only.

Corn Hill Nursery
RR 5
Petitcodiac, New Brunswick E0A 2H0
Hardy elderberries, currants, gooseberries, raspberries, strawberries. Catalogue $2, to Canada only.

Henry Field's Seed & Nursery Co.
Shenandoah, Iowa 51602
Large selection of strawberries and blueberries. Also boysenberries, dewberries, blackberries, raspberries, tayberries, wineberries, currants, gooseberries, serviceberries, mulberries, elderberries. Catalogue free, to U.S. only.

Gurney Seed & Nursery Company
Gurney Building
Yankton, South Dakota 57079
Blackberries, blueberries, boysenberries, currants, elderberries, gooseberries, 'Jostaberries,' mulberries, saskatoons, strawberries, raspberries. Catalogue free.

Hartmann's Plantation Inc.
Box E
310 60th Street
Grand Junction, Michigan 49057
A very good selection of blueberries. Also carry blackberries, raspberries. Catalogue $2.25 (U.S.).

Makielski Berry Farms & Nursery
7130 Platt Road
Ypsilanti, Michigan 48197

Strawberries and raspberries. Catalogue free.

Morden Nurseries
Box 1270
Morden, Manitoba R0G 1J0
Hardy strawberries, raspberries, gooseberries, saskatoons, currants. Catalogue free.

North Star Gardens
19060 Manning Trail North
Marine on St. Croix, Minnesota 55047-9723
An extensive list of raspberries. Catalogue $4 (U.S.).

Nourse Farms, Inc.
Box 458 RFD
South Deerfield, Massachusetts 01373
Blackberries, raspberries, strawberries. Catalogue free.

W.H. Perron
515 boulevard Labelle
Ville de Laval, Quebec H7V 2T3
Currants, gooseberries, blackberries, 'Jostaberries,' raspberries, blueberries, strawberries, hardy roses. Catalogue $3.

Rayner Bros. Inc.
Box 1617
Salisbury, Maryland 21801
Good selection of strawberries, blueberries, raspberries. Also blackberries. Catalogue free, to U.S. only.

Le Reveil de la Nature
RR 1
St-Philibert, Quebec G0M 1X0
Blueberries, strawberries, gooseberries, currants, raspberries. French-language catalogue $1.

St. Lawrence Nurseries
RD 2
Potsdam, New York 13676
Good selection of blueberries. Also elderberries, ground cherries, currants, *Rosa*

rugosa, gooseberries, cranberries, buffalo-berries. Catalogue $1 to U.S., $3 foreign.

La Talle de Framboise
103 Broadview Avenue
Pointe Claire, Quebec H9R 3Z3
Very good selection of blackberries, raspberries, currants, gooseberries. English-language catalogue $2.

Windmill Point Farm & Nursery
2103 Perrot Boulevard
ND Ile Perrot, Quebec J7V 5V6
Gooseberries, currants, elderberries, mulberries, saskatoons, blueberries, cranberries, mountain ash, buffalo-berries, hawthorn, *Rosa rugosa*, raspberries, tayberries, wineberries, strawberries. Catalogue $2.50.

Windy Ridge Nurseries
Box 301
Hythe, Alberta T0H 2C0
Good selection of hardy strawberries, currants, saskatoons, raspberries and gooseberries, as well as buffalo-berries, highbush cranberries and blueberries. Catalogue $1.

Index

Credits

Contents